Forces, Motion and Energy

Interactive Textbook

HOLT, RINEHART AND WINSTON
A Harcourt Education Company
Orlando • **Austin** • New York • San Diego • London

Printed in the United States of America

ISBN 0-03-095824-5

6 7 8 9 0956 12 11 10

Contents

CHAPTER 1 Matter in Motion

CHAPTER 2 Forces and Motion

CHAPTER 3 Forces in Fluids

CHAPTER 4 Work and Machines

CHAPTER 5 Energy and Energy Resources

CHAPTER 6 Heat and Heat Technology

Name ______________________ Class ______________ Date ____________

CHAPTER 1 Matter in Motion

SECTION 1

Measuring Motion

BEFORE YOU READ

After you read this section, you should be able to answer these questions:

- What is motion?
- How is motion shown by a graph?
- What are speed and velocity?
- What is acceleration?

National Science Education Standards
PS 2a

What Is Motion?

Look around the room for a moment. What objects are in motion? Are students writing with pencils in their notebooks? Is the teacher writing on the board? Motion is all around you, even when you can't see it. Blood is circulating throughout your body. Earth orbits around the sun. Air particles shift in the wind.

STUDY TIP

Describe Study each graph carefully. In the margin next to the graph, write a sentence or two explaining what the graph shows.

When you watch an object move, you are watching it in relation to what is around it. Sometimes the objects around the object you are watching are at rest. An object that seems to stay in one place is called a *reference point*. When an object changes position over time in relation to a reference point, the object is in **motion**. ☑

READING CHECK

1. Describe What is the purpose of a reference point?

____

____

You can use *standard reference directions* (such as north, south, east, west, right, and left) to describe an object's motion. You can also use features on Earth's surface, such as buildings or trees, as reference points. The figure below shows how a mountain can be used as a reference point to show the motion of a hot-air balloon.

The hot-air balloon changed position relative to a reference point.

TAKE A LOOK

2. Identify What is the fixed reference point in the photos?

____

How Can Motion Be Shown?

In the figure below, a sign-up sheet is being passed around a classroom. You can follow its path. The paper begins its journey at the reference point, the origin.

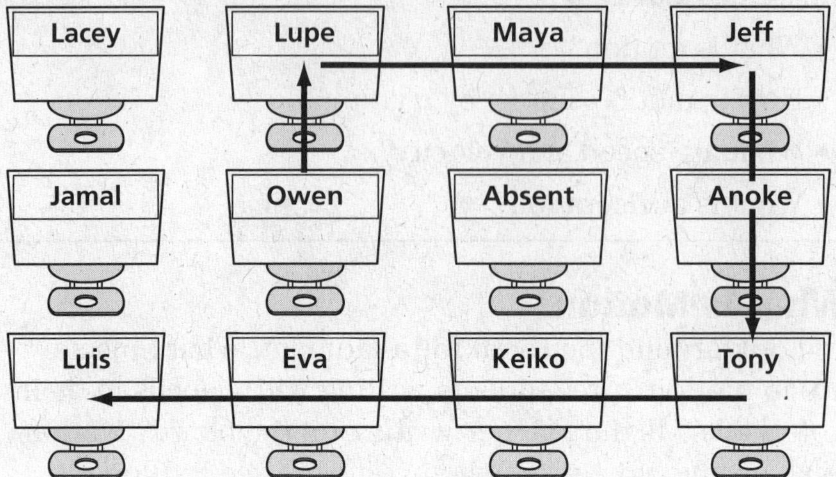

The path taken by a field trip sign-up sheet.

TAKE A LOOK

3. Identify What is the origin, or reference point, of the paper?

____

The figure below shows a graph of the position of the sign-up sheet as it is passed around the class. The paper moves in this order:

1. One positive unit on the y-axis
2. Two positive units on the x-axis
3. Two negative units on the y-axis
4. Three negative units on the x-axis

The graph provides a method of using standard reference directions to show motion.

STANDARDS CHECK

PS 2a The motion of an object can be described by its position, direction of motion, and speed. That motion can be measured and represented on a graph.

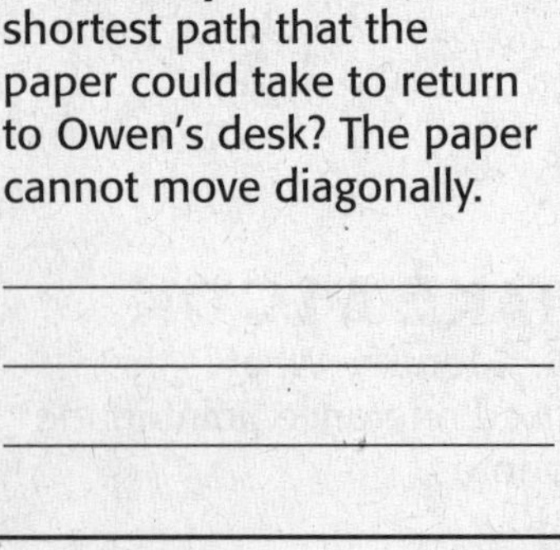

4. Identify What is the shortest path that the paper could take to return to Owen's desk? The paper cannot move diagonally.

____

____

____

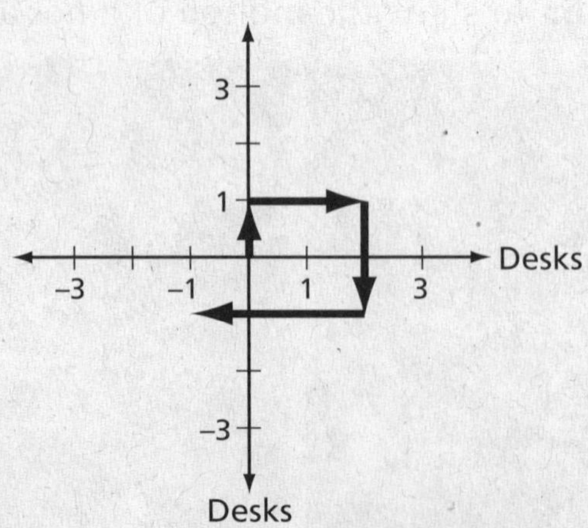

The position of the sign-up sheet as it moves through the classroom.

Name ______________________ Class ______________ Date ______________

What Is Speed?

Speed is the rate at which an object moves. It is the distance traveled divided by the time taken to travel that distance. Most of the time, objects do not travel at a constant speed. For example, when running a race, you might begin slowly but then sprint across the finish line.

So, it is useful to calculate *average speed*. We use the following equation:

$$\textit{average speed} = \frac{\textit{total distance}}{\textit{total time}}$$

Suppose that it takes you 2 s to walk 4 m down a hallway. You can use the equation above to find your average speed:

$$\textit{average speed} = \frac{4 \text{ m}}{2 \text{ s}} = 2 \text{ m/s}$$

Your speed is 2 m/s. Units for speed include meters per second (m/s), kilometers per hour (km/h), feet per second (ft/s), and miles per hour (mi/h).

Critical Thinking

5. Explain The average flight speed of a bald eagle is about 50 km/h. A scientist has measured an eagle flying 80 km/h. How is this possible?

____

____

____

____

Math Focus

6. Calculate Suppose you walk 10 m down a hallway in 2.5 s. What is your average speed? Show your work.

How Can You Show Speed on a Graph?

You can show speed on a graph by showing how the position of an object changes over time. The x-axis shows the time it takes to move from place to place. The y-axis shows distance from the reference point.

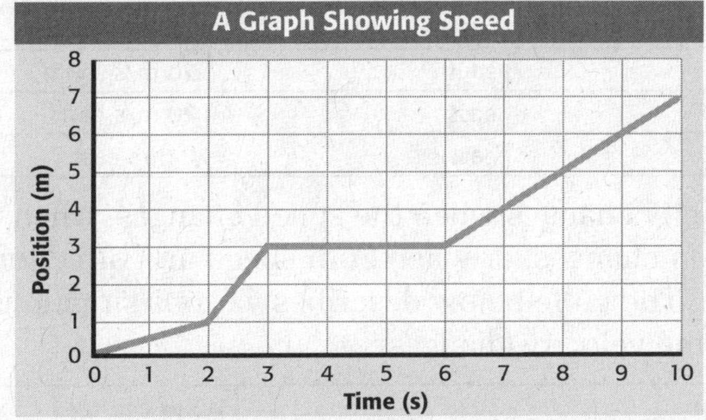

A graph of position versus time also shows the dog's speed during his walk. The more slanted the line, the faster the dog walked.

TAKE A LOOK

7. Apply Concepts Suppose the dog walks at a constant speed the whole way. On the graph, draw a line showing that the dog walks at a constant speed during the walk.

Suppose you watched a dog walk beside a fence. The graph above shows the total distance the dog walked in 10 s. The line is not straight because the dog did not walk the same distance in each second. The dog walked slowly for 2 s and then quickly for 1 s. From 3 s to 5 s, the dog did not move.

The average speed of the dog is:

$$average\ speed = \frac{total\ distance\ walked}{total\ time} = \frac{7\text{ m}}{10\text{ s}} = .07\text{ m/s}$$

What Is Velocity?

Suppose that two birds leave the same tree at the same time. They both fly at 10 km/h for 5 min, then 5 km/h for 10 min. However, they don't end up in the same place. Why not?

The birds did not end up in the same place because they flew in different directions. Their speeds were the same, but because they flew in different directions, their velocities were different. **Velocity** is the speed of an object and its direction. ☑

READING CHECK

8. Analyze Someone tells you that the velocity of a car is 55 mi/h. Is this correct? Explain your answer.

____

____

The velocity of an object is constant as long as both speed and direction are constant. If a bus driving at 15 m/s south speeds up to 20 m/s south, its velocity changes. If the bus keeps moving at the same speed but changes direction from south to east, its velocity also changes. If the bus brakes to a stop, the velocity of the bus changes again.

Say It

Share Experiences Have you ever experienced a change in velocity on an amusement park ride? In pairs, share an experience. Explain how the velocity changed—was it a change in speed, direction, or both?

The table below shows that velocity is a combination of both the speed of an object and its direction.

Speed	Direction	Velocity
15 m/s	south	15 m/s south
20 m/s	south	20 m/s south
20 m/s	east	20 m/s east
0 m/s	east	0 m/s east

Velocity changes when the speed changes, when the direction changes, or when both speed and direction change. The table below describes various situations in which the velocity changes.

TAKE A LOOK

9. Identify Fill in the empty boxes in the table.

Situation	What changes
Raindrop falling faster and faster	
Runner going around a turn on a track	direction
Car taking an exit off a highway	speed and direction
Train arriving at a station	speed
Baseball being caught by a catcher	speed
Baseball hit by a batter	
	speed and direction

This cyclist moves faster and faster as he peddles his bike south.

TAKE A LOOK

10. Identify Is the cyclist accelerating? How do you know?

____

____

What Is Acceleration?

Acceleration is how quickly velocity changes. An object accelerates if its speed changes, its direction changes, or both its speed and direction change.

The units for acceleration are the units for velocity divided by a unit for time. The resulting unit is often meters per second per second (m/s/s or m/s^2).

Looking at the figure above, you can see that the speed increases by 1 m/s during each second. This means that the cyclist is accelerating at 1 m/s^2.

An increase in speed is referred to as *positive acceleration.* A decrease in speed is referred to as *negative acceleration* or *deceleration.* ☑

READING CHECK

11. Explain What happens to an object when it has negative acceleration?

____

____

Acceleration can be shown on a graph of speed versus time. Suppose you are operating a remote control car. You push the lever on the remote to move the car forward. The graph below shows the car's acceleration as the car moves east. For the first 5 s, the car increases in speed. The car's acceleration is positive because the speed increases as time passes.

For the next 2 s, the speed of the car is constant. This means the car is no longer accelerating. Then the speed of the car begins to decrease. The car's acceleration is then negative because the speed decreases over time.

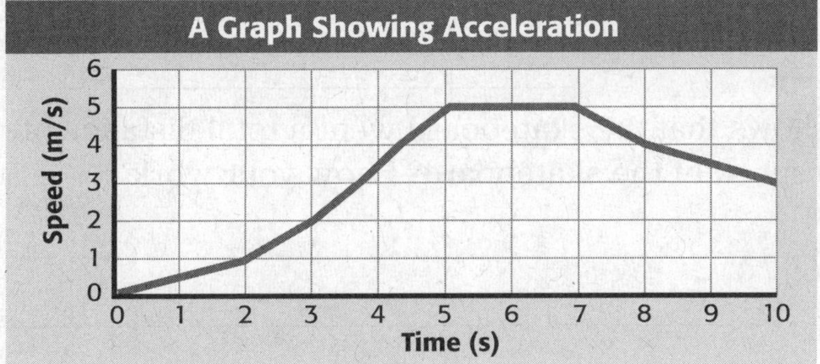

The graph of speed versus time also shows that the acceleration of the car was positive and negative. Between 5 s and 7 s, it had no acceleration.

Math Focus

12. Interpret Graphs Is the slope positive or negative when the car's speed increases? Is the slope positive or negative when the car's speed decreases?

____

____

____

Name ______________________ Class ______________ Date ______________

Section 1 Review

NSES PS 2a

SECTION VOCABULARY

acceleration the rate at which the velocity changes over time; an object accelerates if its speed, direction, or both change **speed** the distance traveled divided by the time interval during which the motion occurred	**motion** an object's change in position relative to a reference point **velocity** the speed of an object in a particular direction

1. Identify What is the difference between speed and velocity?

____

____

____

2. Complete a Graphic Organizer Fill in the graphic organizer for a car that starts from one stop sign and approaches the next stop sign. Use the following terms: constant *velocity*, *positive acceleration*, *deceleration*, and *at rest.*

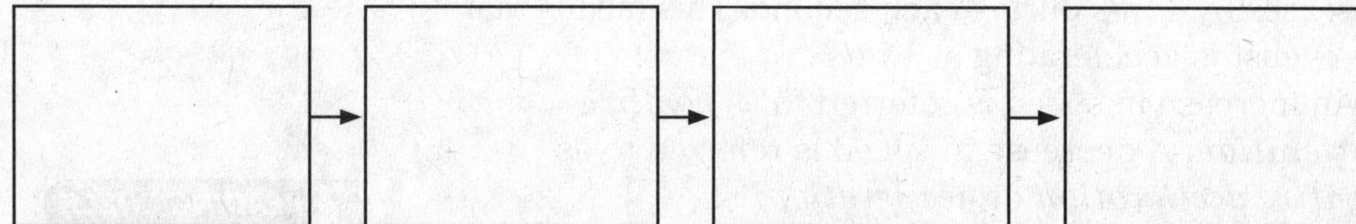

3. Interpret a Graph Describe the motion of the skateboard using the graph below. Write what the skateboard does from time = 0 s to time = 40 s.

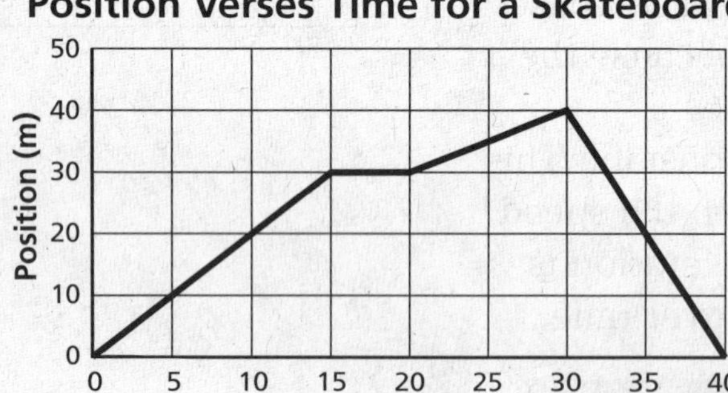

____

____

____

____

____

____

4. Calculate The graph above shows that the skateboard went a total distance of 80 m. What was the average speed of the skateboard? Show your work.

____

____

____

Name ______________________ Class ______________ Date ______________

CHAPTER 1 Matter in Motion

SECTION 2

What Is a Force?

BEFORE YOU READ

After you read this section, you should be able to answer these questions:

- What is a force?
- How do forces combine?
- What is a balanced force?
- What is an unbalanced force?

National Science Education Standards
PS 2b, PS 2c

What Is a Force?

You probably hear people talk about force often. You may hear someone say, "That storm had a lot of force" or "Mrs. Larsen is the force behind the school dance." But what exactly is a force in science?

In science, a **force** is a push or a pull. All forces have two properties: direction and size. A **newton** (N) is the unit that describes the size of a force. ☑

Forces act on the objects around us in ways that we can see. If you kick a ball, the ball receives a push from you. If you drag your backpack across the floor, the backpack is pulled by you.

Forces also act on objects around us in ways that we cannot see. For example, in the figure below, a student is sitting on a chair. What are the forces acting on the chair?

The student is pushing down on the chair, but the chair does not move. Why? The floor is balancing the force by pushing up on the chair. When the forces on an object are *balanced*, the object does not move.

STUDY TIP

Brainstorm As you read, think about objects you see every day. What kinds of forces are affecting them? How do the forces affect them?

READING CHECK

1. List What two properties do all forces have?

____

____

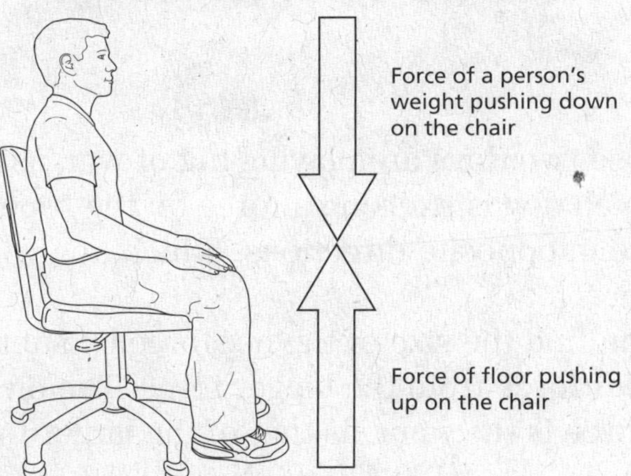

A person sitting on a chair.

TAKE A LOOK

2. Explain Since the chair is not moving, what kind of forces are acting on it?

____

____

How Do Forces Combine?

As you saw in the previous example, more than one force often acts on an object. When all of the forces acting on an object are added together, you determine the **net force** on the object. An object with a net force more than 0 N acting on it will change its state of motion.

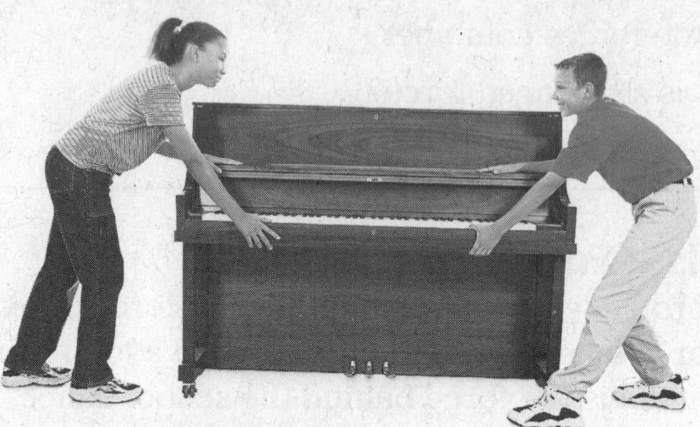

TAKE A LOOK

3. Identify On the figure, draw an arrow showing the direction and size of the net force on the piano. The length of the arrow should represent the size of the force.

FORCES IN THE SAME DIRECTION

Suppose your music teacher asks you and a friend to move a piano, as shown in the figure above. You push the piano from one end and your friend pulls the piano from the other end. You and your friend are applying forces in the same direction. Adding the two forces gives you the size of the net force. The direction of the net force is the same as the direction of the forces.

$$125\text{ N} + 120\text{ N} = 245\text{ N}$$

net force = 245 N to the right

FORCES IN DIFFERENT DIRECTIONS

Critical Thinking

4. Predict What would happen if both dogs pulled the rope with a force of 85 N?

____

____

____

Suppose two dogs are playing tug of war, as shown above. Each dog is exerting a force on the rope. Here, the forces are in opposite directions. Which dog will win the tug of war?

You can find the size of the net force by subtracting the smaller force from the bigger force. The direction of the net force is the same as that of the larger force:

$$120\text{ N} - 80\text{ N} = 40\text{ N}$$

net force = 40 N to the right

What Happens When Forces Are Balanced or Unbalanced?

Knowing the net force on an object lets you determine its effect on the motion of the object. Why? The net force tells you whether the forces on the object are balanced or unbalanced.

BALANCED FORCES

When the forces on an object produce a net force of 0 N, the forces are *balanced*. There is no change in the motion of the object. For example, a light hanging from the ceiling does not move. This is because the force of gravity pulls down on the light while the force of the cord pulls upward. ☑

READING CHECK

5. Describe What happens to the motion of an object if the net force acting on it is 0 N?

____

____

The soccer ball moves because the players exert an unbalanced force on the ball each time they kick it.

UNBALANCED FORCES

When the net force on an object is not 0 N, the forces on the object are *unbalanced*. Unbalanced forces produce a change in motion of an object. Think about a soccer game. Players kick the ball to each other. When a player kicks the ball, the kick is an unbalanced force. It sends the ball in a new direction with a new speed.

An object can continue to move when the unbalanced forces are removed. For example, when it is kicked, a soccer ball receives an unbalanced force. The ball continues to roll on the ground after the ball was kicked until an unbalanced force changes its motion.

STANDARDS CHECK

PS 2c If more than one force acts on an object along a straight line, then the forces reinforce or cancel one another, depending on their direction and magnitude. Unbalanced forces will cause changes in the speed or direction of an object's motion.

6. Describe What will happen to an object that has an unbalanced force acting on it?

____

____

Name ______________ Class ______________ Date ______________

Section 2 Review

NSES PS 2b, PS 2c

SECTION VOCABULARY

force a push or a pull exerted on an object in order to change the motion of the object; force has size and direction **net force** the combination of all the forces acting on an object	**newton** the SI unit for force (symbol, N)

1. **Explain** If there are many forces acting on an object, how can the net force be 0?

2. **Apply Concepts** Identify three forces acting on a bicycle when you ride it.

3. **Calculate** Determine the net force on each of the objects shown below. Don't forget to give the direction of the force.

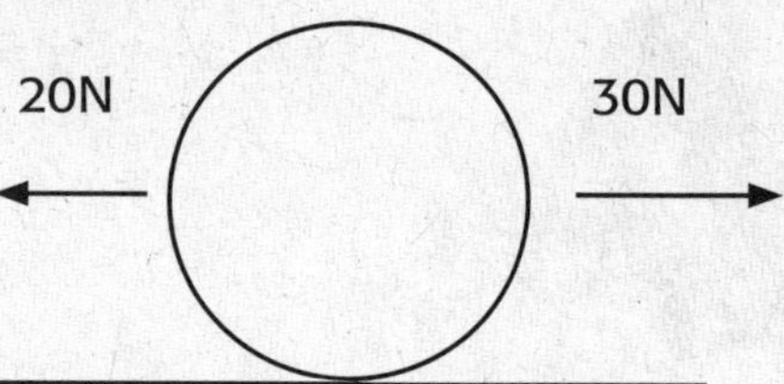

net force = ______________ *net force* = ______________

4. **Explain** How will the net force affect the motion of each object shown above?

5. **Describe** What is the difference between balanced and unbalanced forces?

Name _______________ Class _______________ Date _______________

SECTION 3

Friction: A Force That Opposes Motion

BEFORE YOU READ

After you read this section, you should be able to answer these questions:

- What is friction?
- How does friction affect motion?
- What are the types of friction?
- How can friction be changed?

National Science Education Standards
PS 2c

What Causes Friction?

Suppose you are playing soccer and you kick the ball far from you. You know that the ball will slow down and eventually stop. This means that the velocity of the ball will decrease to 0. You also know that an unbalanced force is needed to change the velocity of objects. So, what force is stopping the ball?

Friction is the force that opposes the motion between two surfaces that touch. Friction causes the ball to slow down and then stop. ☑

What causes friction? The surface of any object is rough. Even an object that feels smooth is covered with very tiny hills and valleys. When two surfaces touch, the hills and valleys of one surface stick to the hills and valleys of the other. This contact between surfaces causes friction.

If a force pushes two surfaces together even harder, the hills and valleys come closer together. This increases the friction between the surfaces.

STUDY TIP

Imagine As you read, think about the ways that friction affects your life. Make a list of things that might happen, or not happen, if friction did not exist.

READING CHECK

1. Describe What is friction?

____

____

____

____

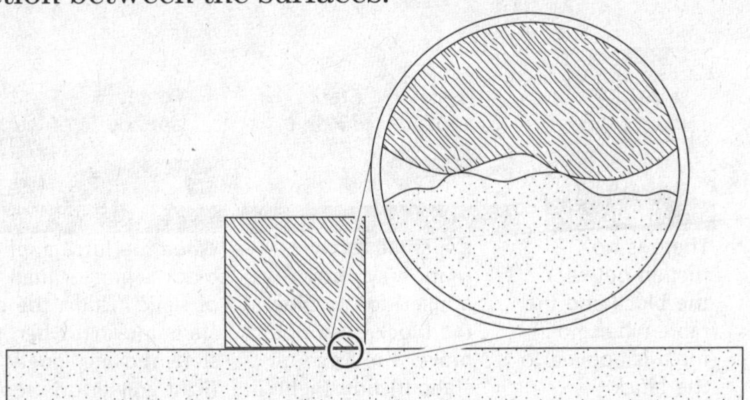

When the hills and valleys of one surface stick to the hills and valleys of another surface, friction is created.

TAKE A LOOK

2. Explain Why can't you see the hills and valleys without a close-up view of the objects?

____

____

STANDARDS CHECK

PS 2c If more than one force acts on an object along a straight line, then the forces reinforce or cancel one another, depending on their direction and magnitude. Unbalanced forces will cause changes in the speed or direction of an object's motion.

3. Identify Two identical balls begin rolling next to each other at the same velocity. One is on a smooth surface and one is on a rough surface. Which ball will stop first? Why?

____

____

____

What Affects the Amount of Friction?

Imagine that a ball is rolled over a carpeted floor and another ball is rolled over a wood floor. Which surface affects a ball more?

The smoothness of the surfaces of the objects affects how much friction exists. Friction is usually greater between materials that have rough surfaces than materials that have smooth surfaces. The carpet has greater friction than the wood floor, so the ball on the carpet stops first.

What Types of Friction Exist?

There are two types of friction: kinetic friction and static friction. *Kinetic friction* occurs when force is applied to an object and the object moves. When a cat slides along a countertop, the friction between the cat and the countertop is kinetic friction. The word kinetic means "moving."

The amount of kinetic friction between moving surfaces depends partly on how the surfaces move. In some cases, the surfaces slide past each other like pushing a box on the floor. In others, one surface rolls over another like a moving car on a road. There is usually less friction between surfaces that roll than between surfaces that slide.

Static friction occurs when force applied to an object does not cause the object to move. When you try to push a piece of furniture that will not move, the friction observed is static friction.

TAKE A LOOK

4. Describe When does static friction become kinetic friction?

____

____

____

____

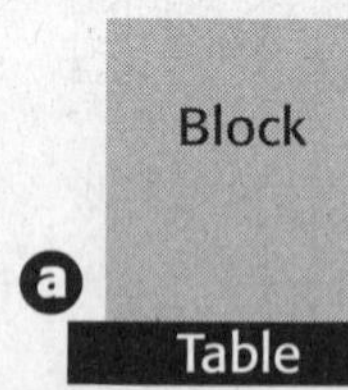

There is no friction between the block and the table when no force is applied to the block.

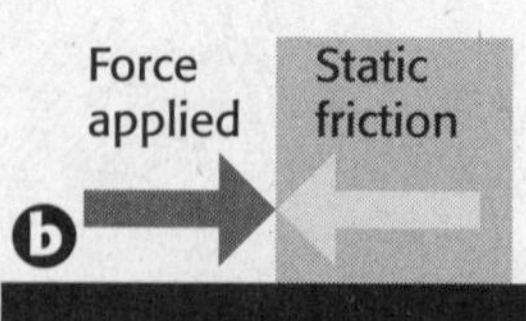

If a small force (dark gray arrow) is applied to the block, the block does not move. The force of static friction (light gray arrow) balances the force applied.

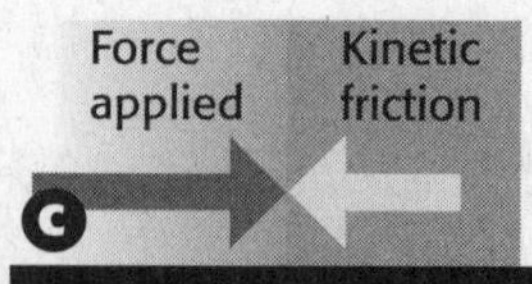

When the force applied to the block is greater than the force of static friction, the block starts moving. When the block starts moving, kinetic friction (light gray arrow) replaces all of the static friction and opposes the force applied.

How Can Friction Be Decreased?

To reduce the amount of friction, you can apply a lubricant between two surfaces. A *lubricant* is a substance that reduces the friction between surfaces. Motor oil, wax, and grease are examples of lubricants.

You can also reduce friction by rolling, rather than sliding, an object. A refrigerator on rollers is much easier to move than one that just slides.

Another way of reducing friction is to smooth the surfaces that rub against each other. Skiers have their skis sanded down to make them smoother. This makes it easier for the skis to slide over the snow.

If you work on a bicycle, you may get dirty from the chain oil. This lubricant reduces friction between sections of the chain.

Critical Thinking

5. Infer How does a lubricant reduce the amount of friction?

______

TAKE A LOOK

6. Identify Why is it important to put oil on a bicycle chain?

______

How Can Friction Be Increased?

Increasing the amount of friction between surfaces can be very important. For example, when the tires of a car grip the road better, the car stops and turns corners much better. Friction causes the tires to grip the road. Without friction, a car could not start moving or stop.

On icy roads, sand can be used to make the road surface rougher. Friction increases as surfaces are made rougher.

You can also increase friction by increasing the force between the two objects. Have you ever cleaned a dirty pan in the kitchen sink? You may have found that cleaning the pan with more force allows you to increase the amount of friction. This makes it easier to clean the pan. ☑

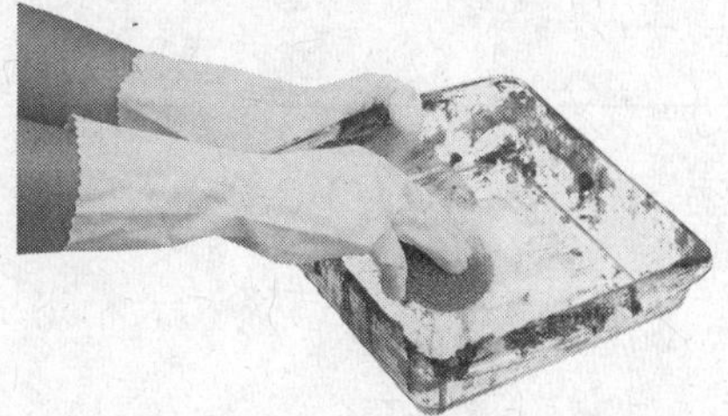

✓ READING CHECK

7. Identify Name two things that can be done to increase the friction between surfaces.

______

Name ______________________ Class ______________ Date ____________

Section 3 Review

NSES PS 2c

SECTION VOCABULARY

friction a force that opposes the motion between two surfaces that are in contact	

1. Describe What effect does friction have when you are trying to move an object at rest?

____

2. Compare Explain the difference between static friction and kinetic friction. Give an example of each.

____

____

____

3. Compare The figure on the left shows two surfaces up close. On the right, draw a sketch. Show what the surfaces of two objects that have less friction between them might look like.

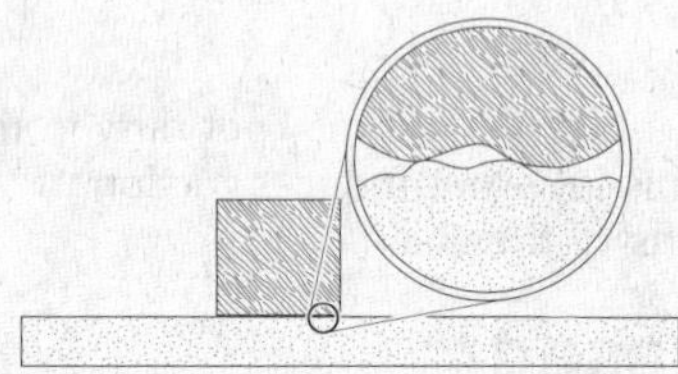

4. Analyze Name three common lubricants and describe why they are used.

____

____

5. Analyze In what direction does friction always act?

____

6. Identify A car is driving on a flat road. When the driver hits the brakes, the car slows down and stops. What would happen if there were no friction between the tires and the road? Explain your answer.

____

____

Name ______________________ Class ______________ Date ______________

CHAPTER 1 Matter in Motion

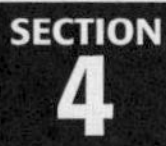

SECTION 4 Gravity: A Force of Attraction

BEFORE YOU READ

After you read this section, you should be able to answer these questions:

- What is gravity?
- How are weight and mass different?

National Science Education Standards
PS 2c

How Does Gravity Affect Matter?

Have you ever seen a video of astronauts on the moon? The astronauts bounce around like beach balls even though the space suits weighed 180 pounds on Earth. See the figure below. Why is it easier for a person to move on the moon than on Earth? The reason is that the moon has less gravity than Earth. **Gravity** is a force of attraction, or a pull, between objects. It is caused by their masses. ☑

All matter has mass. Gravity is a result of mass. Therefore all matter has gravity. This means that all objects attract all other objects in the universe! The force of gravity pulls objects toward each other. For example, gravity between the objects in the solar system holds the solar system together. Gravity holds you to Earth.

Small objects also have gravity. You have gravity. This book has gravity. Why don't you notice the book pulling on you or you pulling on the book? The reason is that the book's mass and your mass are both small. The force of gravity caused by small mass is not large enough to move either you or the book.

Because the moon has less gravity than Earth does, walking on the moon's surface was a very bouncy experience for the Apollo astronauts.

STUDY TIP

Discuss Ideas Take turns reading this section out loud with a partner. Stop to discuss ideas that seem confusing.

READING CHECK

1. Describe What is gravity?

____

____

____

Critical Thinking

2. Infer Why can't you see two soccer balls attracting each other?

____

____

____

What Is the Law of Universal Gravitation?

According to a story, Sir Isaac Newton, while sitting under an apple tree, watched an apple fall. This gave him a bright idea. Newton realized that an unbalanced force on the apple made it fall.

He then thought about the moon's orbit. Like many others, Newton had wondered what kept the planets in the sky. He realized that an unbalanced force on the moon kept it moving around Earth. Newton said that these forces are both gravity.

Newton's ideas are known as the *law of universal gravitation.* Newton said that all objects in the universe attract each other because of gravitational force. ☑

READING CHECK

3. Describe What is the law of universal gravitation?

____

____

____

____

This law says that gravitational force depends on two things:

1. the masses of the objects
2. the distance between the objects

The word "universal" means that the law applies to all objects. See the figure below. ☑

READING CHECK

4. Identify What two things determine gravitational force?

____

____

STANDARDS CHECK

PS 2c If more than one force acts on an object along a straight line, then the forces reinforce or cancel one another, depending on their direction and magnitude. Unbalanced forces will cause changes in the speed or direction of an object's motion.

5. Identify What is the unbalanced force that affects the motions of a falling apple and the moon?

____

Sir Isaac Newton said that the same unbalanced force caused the motions of the apple and the moon.

How Does Mass Affect Gravity?

Imagine an elephant and a cat. Because the elephant has a larger mass than the cat does, gravity between the elephant and Earth is larger. So, the cat is much easier to pick up than the elephant. The gravitational force between objects depends on the masses of the objects. See the figure below.

Gravitational force is small between objects that have small masses.

If the mass of one or both objects increases, the gravitational force pulling them together increases.

The arrows indicate the gravitational force between two objects. The length of the arrows indicates the magnitude of the force.

TAKE A LOOK

6. Compare Is there more gravitational force between objects with small masses or objects with large masses?

____

____

____

Mass also explains why an astronaut on the moon can jump around so easily. The moon has less mass than Earth does. This gives the moon a weaker pull on objects than the pull of Earth. The astronaut is not being pulled toward the moon as much as he is by Earth. So the astronaut can jump higher and more easily on the moon.

The universal law of gravitation can let us predict what happens to gravity when mass changes. According to the universal law of gravitation, suppose there is a 5 N force of gravity between two objects. If the mass of one object doubles and the other stays the same, the force of gravity also doubles.

Let's try a problem. The force due to gravity between two objects is 3 N. If the mass of one object triples and the other stays the same, what is the new force of gravity?

Solution: Since the mass of one object tripled and the other stayed the same, the force of gravity also triples. It is 9 N.

Math Focus

7. Infer Two objects of equal mass have a force of gravity of 6 N between them. Imagine the mass of one is cut in half and the other stays the same, what is the force due to gravity?

____

How Does Distance Affect Gravity?

The mass of the sun is 300,000 times bigger than that of Earth. However, if you jump up, you return to Earth every time you jump rather than flying toward the sun. If the sun has more mass, then why doesn't it have a larger gravitational pull on you?

This is because the gravitational force also depends on the distance between the objects. As the distance between two objects gets larger, the force of gravity gets much smaller. And as the distance between objects gets smaller, the force of gravity gets much bigger. This is shown in the figure below.

Although the sun has tremendous mass, it is also very far away. This means that it has very little gravitational force on your body or on small objects around you. The sun does have a large gravitational force on planets because the masses of planets are very large.

Critical Thinking

8. Analyze The sun is much more massive than Earth. Why is the force of gravity between you and the sun so much less than Earth's gravity and you?

____

____

Gravitational force is large when the distance between two objects is small.

TAKE A LOOK

9. Describe Use the diagram to describe the effect of distance on gravitational force.

____

____

____

____

____

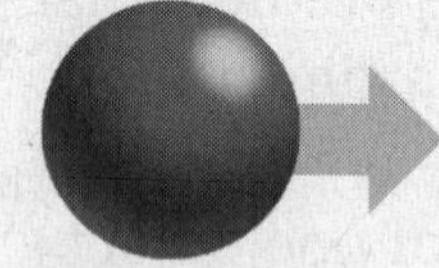

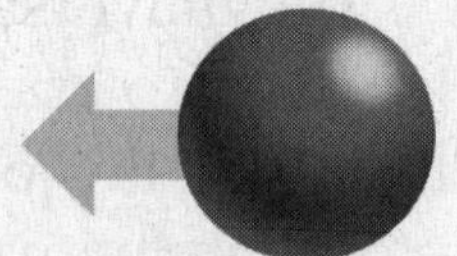

If the distance between two objects increases, the gravitational force pulling them together decreases rapidly.

The length of the arrows indicates the magnitude of the gravitational force between two objects.

What Is the Difference Between Mass and Weight?

You have learned that gravity is a force of attraction between objects. **Weight** is a measure of the gravitational force on an object. The SI unit for weight is the newton (N).

Mass is a measure of the amount of matter in an object. This seems similar to weight, but it is not the same. An object's mass does not change when gravitational forces change, but its weight does. Mass is usually expressed in kilograms (kg) or grams (g). ☑

In the figure below, you can see the difference between mass and weight. Compare the astronaut's mass and weight on Earth to his mass and weight on the moon.

READING CHECK

10. Contrast How is mass different from weight?

____

____

____

____

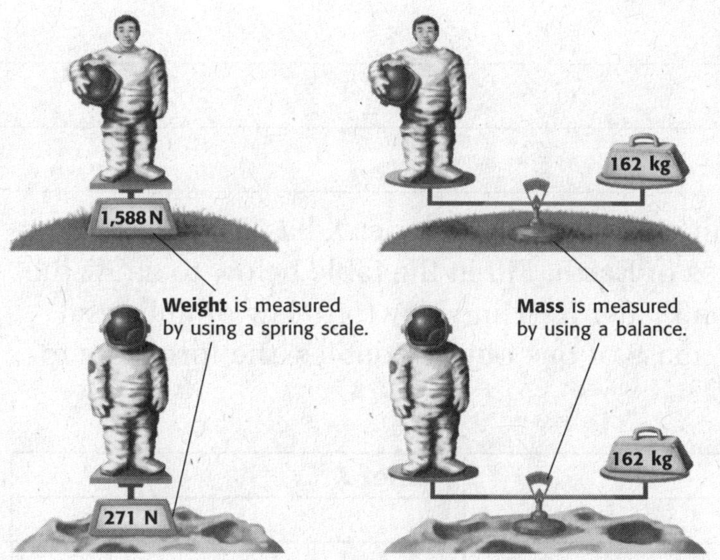

TAKE A LOOK

11. Identify What is the weight of the astronaut on Earth? What is the weight of the astronaut on the moon?

____

____

Gravity can cause objects to move because it is a type of force. But gravity also acts on objects that are not moving, or *static*. Earth's gravity pulls objects downward. However, not all objects move downward. Suppose a framed picture hangs from a wire. Gravity pulls the picture downward, but tension (the force in the wire) pulls the picture upward. The forces are balanced so that framed picture does not move.

Critical Thinking

12. Contrast What forces act on a framed picture on a shelf?

____

____

Name ______________________ Class ______________ Date ______________

Section 4 Review

NSES PS 2c

SECTION VOCABULARY

gravity a force of attraction between objects that is due to their masses **mass** a measure of the amount of matter in an object	**weight** a measure of the gravitational force exerted on an object; its value can change with the location of the object in the universe

1. Identify What is gravity? What determines the gravitational force between objects?

____

____

2. Describe A spacecraft is moving toward Mars. Its rocket engines are turned off. As the spacecraft nears the planet, what will happen to the pull of Mars's gravity?

____

3. Summarize An astronaut travels from Earth to the moon. How does his mass change? How does his weight change? Explain.

____

____

____

4. Applying Concepts An astronaut visits Planet X. Planet X has the same radius as Earth but has twice the mass of Earth. Fill in the table below to show the astronaut's mass and weight on Planet X. (Hint: Newton's law of universal gravitation says that when the mass of one object doubles, the force due to gravity also doubles.)

	Earth	Planet X
Mass of astronaut	80 kg	
Weight of astronaut	784 N	

5. Select Each of the spheres shown below is made of iron. Circle the pair of spheres that would have the greatest gravitational force between them. Below the spheres, explain the reason for your choice.

 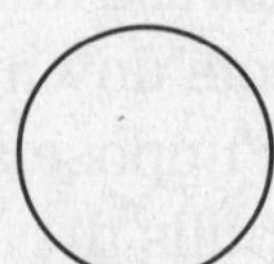   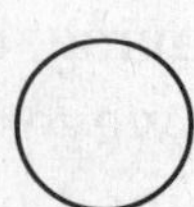

____

____

Name ______________________ Class ______________ Date ______________

CHAPTER 2 Forces and Motion

SECTION 1

Gravity and Motion

BEFORE YOU READ

After you read this section, you should be able to answer these questions:

- How does gravity affect objects?
- How does air resistance affect falling objects?
- What is free fall?
- Why does an object that is thrown horizontally follow a curved path?

How Does Gravity Affect Falling Objects?

In ancient Greece, a great thinker named Aristotle said that heavy objects fall faster than light objects. For almost 2,000 years, people thought this was true. Then, in the late 1500s, an Italian scientist named Galileo Galilei proved that heavy and light objects actually fall at the same rate.

It has been said that Galileo proved this by dropping two cannonballs from the top of a tower at the same time. The cannonballs were the same size, but one was much heavier than the other. The people watching saw both cannonballs hit the ground at the same time. ☑

Why don't heavy objects fall faster than light objects? Gravity pulls on heavy objects more than it pulls on light objects. However, heavy objects are harder to move than light objects. So, the extra force from gravity on the heavy object is balanced by how much harder it is to move.

Falling Balls

Time 1 — Golf ball — Table tennis ball

Time 2

Time 3

Time 4

Practice After every page, stop reading and think about what you've read. Try to think of examples from everyday life. Don't go on to the next section until you think you understand.

READING CHECK

1. Describe What did the people watching the cannonballs see that told them the cannonballs fell at the same rate?

____

____

____

TAKE A LOOK

2. Predict The golf ball is heavier than the table tennis ball. On the figure, draw three circles to show where the golf ball will be at times 2, 3, and 4.

How Much Acceleration Does Gravity Cause?

Because of gravity, all objects accelerate, or speed up, toward Earth at a rate of 9.8 meters per second per second. This is written as 9.8 m/s/s or 9.8 m/s^2. So, for every second an object falls, its velocity (speed) increases by 9.8 m/s. This is shown in the figure below.

Math Focus

3. Calculate How fast is the ball moving at the end of the third second? Explain your answer.

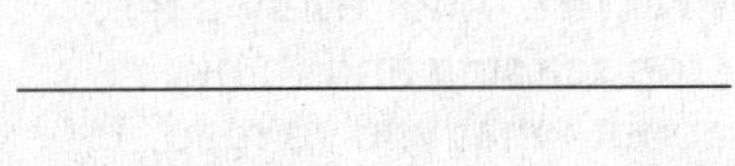

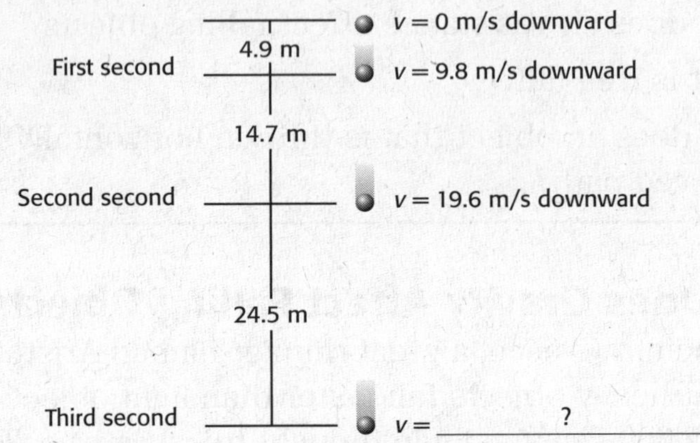

A falling object accelerates at a constant rate. The object falls faster and farther each second than it did the second before.

What Is the Velocity of a Falling Object?

Suppose you drop a rock from a cliff. How fast is it going when it reaches the bottom? If you have a stopwatch, you can calculate its final velocity.

If an object starts from rest and you know how long it falls, you can calculate its final velocity by using this equation:

$$v_{final} = g \times t$$

In the equation, v_{final} stands for final velocity in meters per second, g stands for the acceleration due to gravity (9.8 m/s^2), and t stands for the time the object has been falling (in seconds).

If the rock took 4 s to hit the ground, how fast was it falling when it hit the ground?

Math Focus

4. Calculate A penny is dropped from the top of a tall stairwell. What is the velocity of the penny after it has fallen for 2 s? Show your work.

Step 1: Write the equation.

$$v_{final} = g \times t$$

Step 2: Place values into the equation, and solve for the answer.

$$v_{final} = 9.8 \frac{m/s}{\cancel{s}} \times 4 \cancel{s} = 39.2 \text{ m/s}$$

The velocity of the rock was 39.2 m/s when it hit the ground.

How Can You Calculate How Long an Object Was Falling?

Suppose some workers are building a bridge. One of them drops a metal bolt from the top of the bridge. When the bolt hits the ground, it is moving 49 m/s. How long does it take the bolt to fall to the ground?

Step 1: Write the equation.

$$t = \frac{v_{final}}{g}$$

Step 2: Place values into the equation, and solve for the answer.

$$t = \frac{49\ \cancel{m/s}}{9.8\ \frac{\cancel{m/s}}{s}} = 5\ s$$

The bolt fell for 5 s before it hit the ground.

Math Focus

5. Calculate A rock falls from a cliff and hits the ground with a velocity of 98 m/s. How long does the rock fall? Show your work.

How Does Air Resistance Affect Falling Objects?

Try dropping a pencil and a piece of paper from the same height. What happens? Does this simple experiment show what you just learned about falling objects? Now crumple the paper into a tight ball. Drop the crumpled paper and the pencil from the same height.

What happens? The flat paper falls more slowly than the crumpled paper because of air resistance. *Air resistance* is the force that opposes the motion of falling objects. ☑

How much air resistance will affect an object depends on the size, shape, and speed of the object. The flat paper has more surface area than the crumpled sheet. This causes the flat paper to fall more slowly.

6. Identify Which has more air resistance, the flat paper or the crumpled paper?

____

How Air Resistance Affects Velocity

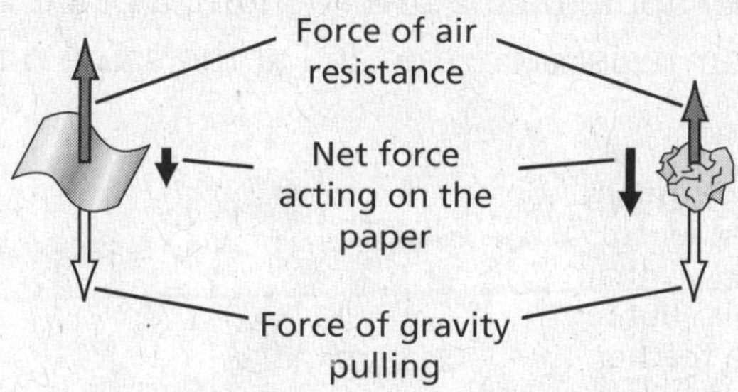

TAKE A LOOK

7. Explain Why does the crumpled paper fall faster than the flat paper?

____

____

____

What Is Terminal Velocity?

As the speed of a falling body increases, air resistance also increases. The upward force of air resistance keeps increasing until it is equal to the downward force of gravity. At this point, the total force on the object is zero, so the object stops accelerating.

When the object stops accelerating, it does not stop moving. It falls without speeding up or slowing down. It falls at a constant velocity called the terminal velocity. **Terminal velocity** is the speed of an object when the force of air resistance equals the force of gravity. ☑

Air resistance causes the terminal velocity of hailstones to be between 5 m/s and 40 m/s. Without air resistance, they could reach the ground at a velocity of 350 m/s! Air resistance also slows sky divers to a safe landing velocity.

READING CHECK

8. Describe When does an object reach its terminal velocity?

____

____

TAKE A LOOK

9. Identify A sky diver is falling at terminal velocity. Draw and label an arrow showing the direction and size of the force due to gravity on the sky diver. Draw and label a second arrow showing the direction and size of the force of air resistance on the sky diver.

The parachute increases the air resistance of this sky diver and slows him to a safe terminal velocity.

What Is Free Fall?

Free fall is the motion of an object when gravity is the only force acting on the object. The figure below shows a feather and an apple falling in a vacuum, a place without any air. Without air resistance, they fall at the same rate.

Air resistance usually causes a feather to fall more slowly than an apple falls. But in a vacuum, a feather and an apple fall with the same acceleration because both are in free fall.

TAKE A LOOK

10. Predict When air resistance acts on the apple and the feather, which falls faster?

____

Orbiting Objects Are in Free Fall

Satellites and the space shuttle orbit Earth. You may have seen that astronauts inside the shuttle float unless they are belted to a seat. They seem weightless. In fact, they are not weightless, because they still have mass.

Weight is a measure of the pull of gravity on an object. Gravity acts between any two objects in the universe. Every object in the universe pulls on every other object. Every object with mass has weight. ☑

The force of gravity between two objects depends on the masses of the objects and how far apart they are. The more massive the objects are, the greater the force is. The closer the objects, the greater the force.

Your weight is determined mostly by the mass of Earth because it is so big and so close to you. If you were to move away from Earth, you would weigh less. However, you would always be attracted to Earth and to other objects, so you would always have weight.

Astronauts float in the shuttle because the shuttle is in free fall. That's right—the shuttle is always falling. Because the astronauts are in the shuttle, they are also falling. The astronauts and the shuttle are falling at the same rate. That is why the astronauts seem to float inside the shuttle. ☑

Isaac Newton first predicted this kind of free fall in the late 17th century. He reasoned that if a cannon were placed on a mountain and fired, the cannon ball would fall to Earth. Yet, if the cannon ball were shot with enough force, it would fall at the same rate that Earth's surface curves away. The cannon ball would never hit the ground, so it would orbit Earth. The figure below shows this "thought experiment."

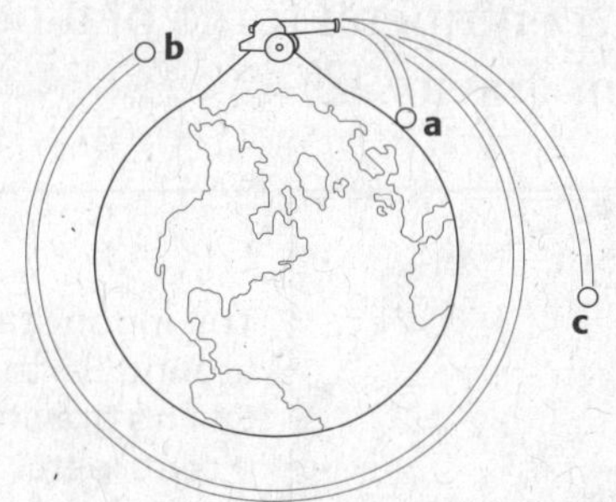

Newton's cannon is a "thought experiment." Newton reasoned that a cannon ball shot hard enough from a mountain top would orbit Earth.

READING CHECK

11. Explain When will an object have no weight? Explain your answer.

____

____

READING CHECK

12. Explain Why don't the astronauts in the orbiting shuttle fall to the floor?

____

____

____

Critical Thinking

13. Infer Compared with cannon ball **b**, what do you think cannon ball **c** would do?

____

____

What Motions Combine to Make an Object Orbit?

An object is in orbit (is orbiting) when it is going around another object in space. When the space shuttle orbits Earth, it is moving forward. Yet, the shuttle is also in free fall. The figure below shows how these two motions combine to cause orbiting.

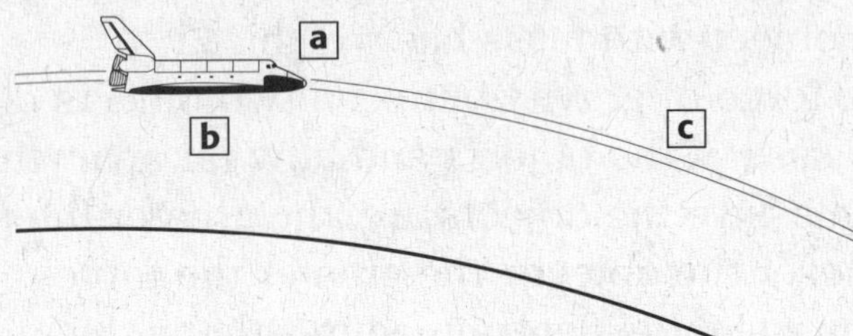

a. The space shuttle moves forward at a constant speed. If there were no gravity, the space shuttle would continue to move in a straight line.

b. The space shuttle is in free fall because gravity pulls it toward Earth. The space shuttle would move straight down if it were not traveling forward.

c. The path of the space shuttle follows the curve of Earth's surface. This path is known as an orbit.

TAKE A LOOK

14. Identify On the figure, draw a line showing the path that the space shuttle would take if gravity were not acting on it.

What Force Keeps an Object in Orbit?

Many objects in space are orbiting other objects. The moon orbits Earth, while Earth and the other planets orbit the sun. These objects all follow nearly circular paths. An object that travels in a circle is always changing direction.

If all the forces acting on an object balance each other out, the object will move in the same direction at the same speed forever. So, objects cannot orbit unless there is an unbalanced force acting on them. *Centripetal force* is the force that keeps an object moving in a circular path. Centripetal force pulls the object toward the center of the circle. The centripetal force of a body orbiting in space comes from gravity. ☑

15. Identify What must be applied to an object to change its direction?

____

TAKE A LOOK

16. Identify Draw an arrow on the figure to show the direction that centripetal force acts on the moon.

The moon stays in orbit around Earth because Earth's gravity provides a centripetal force on the moon.

What Is Projectile Motion?

Projectile motion is the curved path an object follows when it is thrown near the Earth's surface. The motion of a ball that has been thrown forward is an example of projectile motion.

Projectile motion is made up of two parts: horizontal motion and vertical motion. Horizontal motion is motion that is parallel to the ground. Vertical motion is motion that is perpendicular to the ground. The two motions do not affect each other. Instead, they combine to form the curved path we call projectile motion. ☑

When you throw a ball forward, your hand pushes the ball to make it move forward. This force gives the ball its horizontal motion. After the ball leaves your hand, no horizontal forces act on the ball (if we forget air resistance for now). So the ball's horizontal velocity does not change after it leaves your hand.

However, gravity affects the vertical part of projectile motion. Gravity pulls the ball straight down. All objects that are thrown accelerate downward because of gravity.

READING CHECK

17. List What two motions combine to make projectile motion?

____

____

Critical Thinking

18. Infer If you are playing darts and you want to hit the bulls-eye, where should you aim?

____

____

a After the ball leaves the pitcher's hand, the ball's ____________ velocity is constant.

b The ball's vertical velocity increases because ____________ causes it to accelerate downward.

c The two motions combine to form a ____________ path.

TAKE A LOOK

19. List On the figure, fill in the three blanks with the correct words.

20. Apply Concepts If there were no air resistance, how fast would the ball's downward velocity be changing? Explain your answer.

____

____

____

____

Name ______________________ Class ______________ Date ______________

Section 1 Review

SECTION VOCABULARY

free fall the motion of a body when only the force of gravity is acting on the body

projectile motion the curved path that an object follows when thrown, launched, or otherwise projected near the surface of Earth

terminal velocity the constant velocity of a falling object when the force of air resistance is equal in magnitude and opposite in direction to the force of gravity

1. Explain Is a parachutist in free fall? Why or why not?

____

____

____

2. Identify Cause and Effect Complete the table below to show how forces affect objects.

Cause	Effect
Gravity acts on a falling object.	
	The falling object reaches terminal velocity.

3. Calculate A rock at rest falls off a cliff and hits the ground after 3.5 s. What is the rock's velocity just before it hits the ground? Show your work.

4. Identify What force must be applied to an object to keep it moving in a circular path?

____

5. Explain Which part of projectile motion is affected by gravity? Explain how it is affected.

____

____

____

Name ______________________ Class ______________ Date ____________

CHAPTER 2 Forces and Motion

SECTION 2

Newton's Laws of Motion

BEFORE YOU READ

After you read this section, you should, be able to answer these questions:

- What is net force?
- What happens to objects that have no net force acting on them?
- How are mass, force, and acceleration related?
- How are force pairs related by Newton's third law of motion?

National Science Education Standards
PS 2b, 2c

What Is a Net Force?

A *force* is a push or a pull. It is something that causes an object to change speed or direction. There are forces acting on all objects every second of the day. They are acting in all directions.

At first, this might not make sense. After all, there are many objects that are not moving. Are forces acting on an apple sitting on a desk? The answer is yes. Gravity is pulling the apple down. The desk is pushing the apple up.

In the figure below, the arrows represent the size and direction of the forces on the apple.

STUDY TIP

Summarize in Pairs Read each of Newton's laws silently to yourself. After reading each law, talk about what you read with a partner. Together, try to figure out any ideas that you didn't understand.

Forces Acting on an Apple

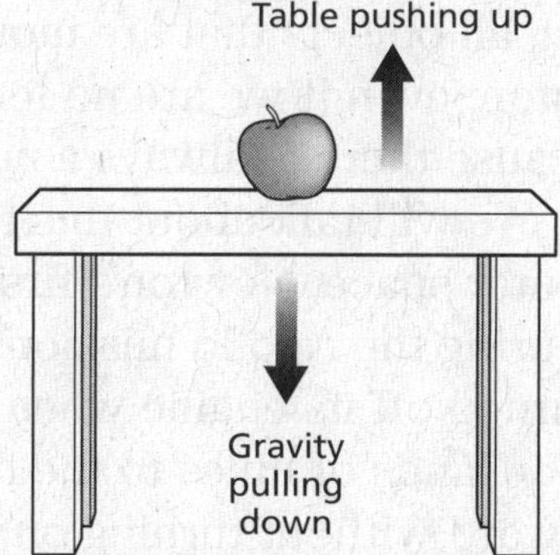

TAKE A LOOK

1. Compare What is the size of the force pulling the apple down compared with the size of the force pushing it up?

____

So, why doesn't the apple move? The apple is staying where it is because all the forces balance out. There are no unbalanced forces. That is, there is no net force on the apple. *Net force* is the total force acting on an object. If the net force on an object is zero, the object will not change speed or direction. ☑

READING CHECK

2. Explain What will not happen to an object if the net force acting on it is zero?

____

____

What Is Newton's First Law of Motion?

Newton's first law of motion describes objects that have no unbalanced forces, or no net force, acting on them. It has two parts:

1. An object at rest will remain at rest.
2. An object moving at a constant velocity will continue to move at a constant velocity. ☑

READING CHECK

3. Identify What will happen to an object at rest if no unbalanced forces act on it?

____

____

4. Identify What will happen to an object moving at constant velocity?

____

____

PART 1: OBJECTS AT REST

An object that is not moving is said to be at rest. A golf ball on a tee is an example of an object at rest. An object at rest will not move unless an unbalanced force is applied to it. The golf ball will keep sitting on the tee until it is struck by a golf club.

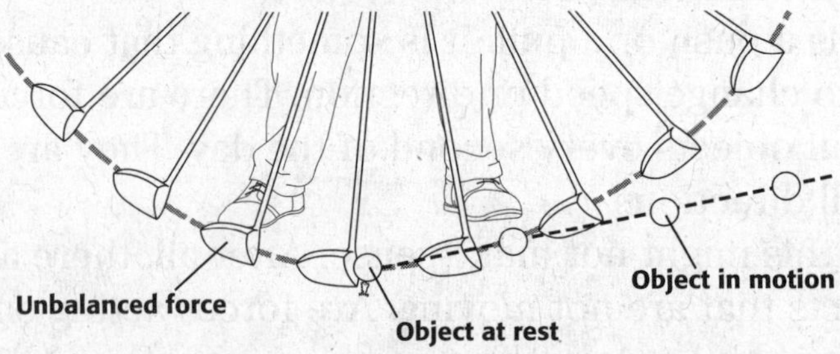

A golf ball will remain at rest on a tee until it is acted on by the unbalanced force of a moving club.

TAKE A LOOK

5. Predict What would happen to the distances between the moving-ball images if the unbalanced force were greater?

____

____

PART 2: OBJECTS IN MOTION

The second part of Newton's first law can be hard to picture. On Earth, all objects that are moving eventually slow down and stop, even if we are no longer touching them. This is because there is always a net force acting on these objects. We will talk about this force later.

However, in outer space, Newton's first law can be seen easily. During the Apollo missions to the moon, the spacecraft turned off its engine when it was in space. It then drifted thousands of miles to the moon. It could keep moving forward without turning on its engines because there was no unbalanced force to slow it down.

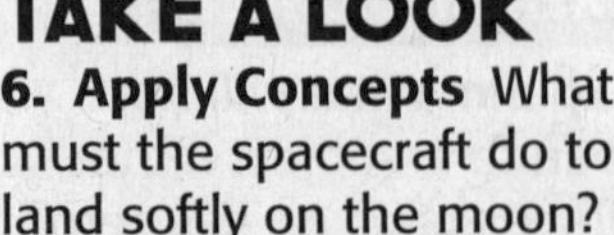

TAKE A LOOK

6. Apply Concepts What must the spacecraft do to land softly on the moon?

____

____

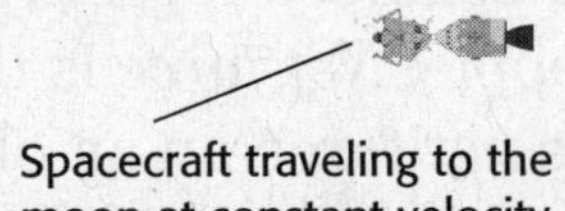

Spacecraft traveling to the moon at constant velocity

How Does Friction Affect Newton's First Law?

On Earth, friction makes observing Newton's first law difficult. If there were no friction, a ball would roll forever until something got in its way. Instead, it stops quickly because of friction.

Friction is a force that is produced whenever two surfaces touch each other. Friction always works against motion. ☑

Friction makes a rolling ball slow down and stop. It also makes a car slow down when its driver lets up on the gas pedal.

READING CHECK

7. Describe How does friction affect the forward motion of an object?

____

What Is Inertia?

Newton's first law is often called the *law of inertia.* **Inertia** is the ability of an object to resist any change in motion. In order to change an object's motion, a force has to overcome the object's inertia. So, in order to move an object that is not moving, you have to apply a force to it. Likewise, in order to change the motion of an object that is moving, you have to apply a force to it. The greater the object's inertia, the harder it is to change its motion.

STANDARDS CHECK

PS 2b An object that is not subjected to a force will continue to move at constant speed and in a straight line.

Word Help: constant
a quantity whose value does not change

8. Explain When a moving car stops suddenly, why does a bag of groceries on the passenger seat fly forward into the dashboard?

____

____

____

____

How Are Mass and Inertia Related?

An object that has a small mass has less inertia than an object with a large mass. Imagine a golf ball and a bowling ball. Which one is easier to move?

The golf ball has much less mass than the bowling ball. The golf ball also has much less inertia. This means that a golf ball will be much easier to move than a bowling ball. ☑

Inertia makes it harder to accelerate a car than to accelerate a bicycle. Inertia also makes it easier to stop a moving bicycle than a car moving at the same speed.

9. Explain Why is a golf ball easier to throw than a bowling ball?

____

____

What Is Newton's Second Law of Motion?

Newton's second law of motion describes how an object moves when an unbalanced force acts on it. The second law has two parts:

1. The acceleration of an object depends on the mass of the object. If two objects are pushed or pulled by the same force, the object with the smaller mass will accelerate more. ☑
2. The acceleration of an object depends on the force applied to the object. If two objects have the same mass, the one you push harder will accelerate more.

READING CHECK

10. Apply Concepts Which object will accelerate more if the same force is applied to both: a pickup truck or a tractor-trailer truck?

____

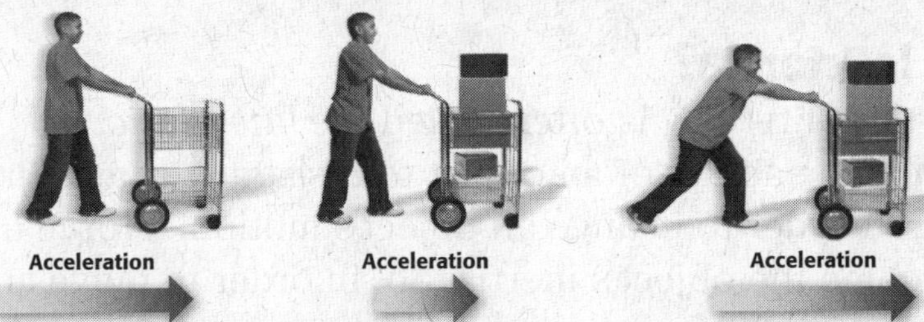

If the force applied to the carts is the same, the acceleration of the empty cart is greater than the acceleration of the loaded cart.

Acceleration increases when a larger force is exerted.

TAKE A LOOK

11. Compare On the figure, draw arrows showing the size and direction of the force that the person is applying to the cart in each picture.

How Is Newton's Second Law Written as an Equation?

Newton's second law can be written as an equation. The equation shows how acceleration, mass, and net force are related to each other:

$$a = \frac{F}{m}, \text{ or } F = m \times a$$

In the equation, a is acceleration (in meters per second squared), m is mass (in kilograms), and F is net force (in newtons, N). One newton is equal to one kilogram multiplied by one meter per second squared. ☑

READING CHECK

12. Identify What are the units of force?

____

Newton's second law explains why all objects fall to Earth with the same acceleration. In the figure on the top of the next page, you can see how the larger force of gravity on the watermelon is balanced by its large mass.

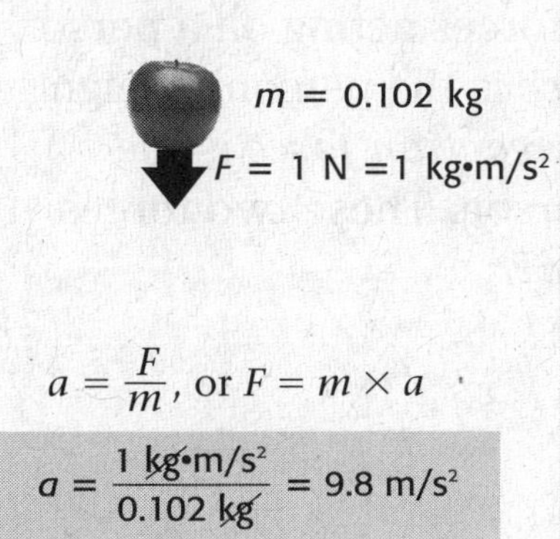

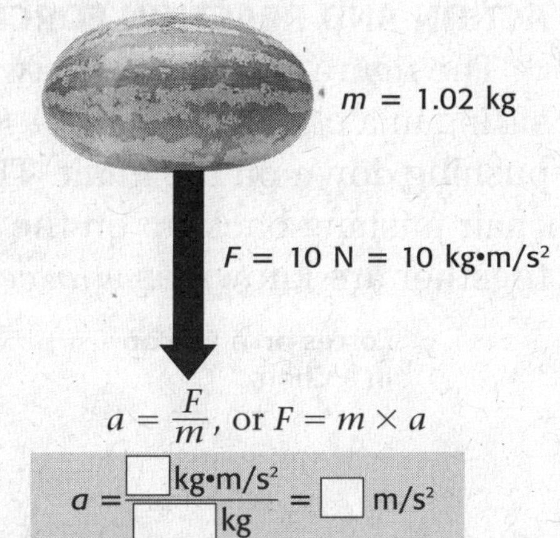

$a = \frac{F}{m}$, or $F = m \times a$

$a = \frac{1 \text{ kg•m/s}^2}{0.102 \text{ kg}} = 9.8 \text{ m/s}^2$

$a = \frac{F}{m}$, or $F = m \times a$

$a = \frac{\square \text{ kg•m/s}^2}{\square \text{ kg}} = \square \text{ m/s}^2$

The apple has less mass than the watermelon does. So, less force is needed to give the apple the same acceleration that the watermelon has.

Math Focus

13. Calculate In the figure, fill in the boxes with the correct numbers to calculate the acceleration of the watermelon.

How Can You Solve Problems Using Newton's Second Law?

You can use the equation $F = ma$ to calculate how much force you need to make a certain object accelerate a certain amount. Or you can use the equation $a = \frac{F}{m}$ to calculate how much an object will accelerate if a certain net force acts on it.

For example, what is the acceleration of a 3 kg mass if a force of 14.4 N is used to move the mass?

Step 1: Write the equation that you will use.

$$a = \frac{F}{m}$$

Step 2: Replace the letters in the equation with the values from the problem.

$$a = \frac{14.4 \text{ N}}{3 \text{ kg}} = \frac{14.4 \text{ kg•m/s}^2}{3 \text{ kg}} = 4.8 \text{ m/s}^2$$

Math Focus

14. Calculate What force is needed to accelerate a 1,250 kg car at a rate of 40 m/s²? Show your work in the space below.

What Is Newton's Third Law of Motion?

All forces act in pairs. Whenever one object exerts a force on a second object, the second object exerts a force on the first object. The forces are always equal in size and opposite in direction.

For example, when you sit on a chair, the force of your weight pushes down on the chair. At the same time, the chair pushes up on you with a force equal to your weight.

ACTION AND REACTION FORCES

The figure below shows two forces acting on a person sitting in a chair. The *action force* is the person's weight pushing down on the chair. The *reaction force* is the chair pushing back up on the person. These two forces together are known as a *force pair*.

Forces on a Person in a Chair

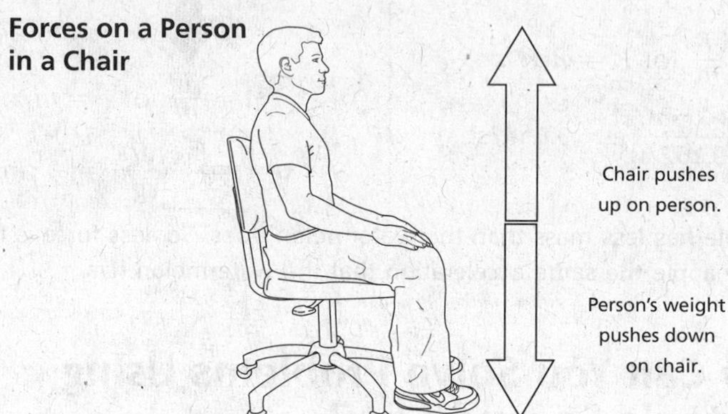

TAKE A LOOK

15. Identify On the figure, label the action force and the reaction force.

Action and reaction forces are also present when there is motion. The figures below show some more examples of action and reaction forces.

TAKE A LOOK

16. Describe How big is the reaction force compared with the action force in each picture?

____

The space shuttle's thrusters push gases downward. The gases push the space shuttle upward with equal force.

The bat exerts a force on the ball and sends the ball flying. The ball exerts an equal force on the bat.

Critical Thinking

17. Apply Concepts If the ball exerts a force on the bat, why doesn't the bat move backward?

____

____

____

____

The action force and reaction force always act on different objects. For example, when you sit in a chair, the action force (your weight) acts on the chair. However, the reaction force (the chair pushing up on you) acts on you.

HARD-TO-SEE REACTION FORCES

In the figure below, a ball is falling toward the Earth's surface. The action force is the Earth's gravity pulling down on the ball. What is the reaction force?

The force of gravity between Earth and a falling object is a force pair.

TAKE A LOOK

18. Identify On the figure, draw and label arrows showing the size and direction of the action force and the reaction force for the ball falling to the Earth.

Believe it or not, the reaction force is the ball pulling up on Earth. Have you ever felt this reaction force when you have dropped a ball? Of course not. However, both forces are present. So, why don't you see or feel Earth rise?

To answer this question, recall Newton's second law. Acceleration depends on the mass and the force on an object. The force acting on Earth is the same as the force acting on the ball. However, Earth has a very, very large mass. Because it has such a large mass, its acceleration is too small to see or feel. ☑

You can easily see the ball's acceleration because its mass is small compared with Earth's mass. Most of the objects that fall toward Earth's surface are much less massive than Earth. This means that you will probably never feel the effects of the reaction force when an object falls to Earth.

READING CHECK

19. Explain Why can't you feel the effect of the reaction force when an object falls to Earth?

____

____

____

Name ______________________ Class ______________ Date ____________

Section 2 Review

NSES PS 2b, 2c

SECTION VOCABULARY

inertia the tendency of an object to resist being moved or, if the object is moving, to resist a change in speed or direction until an outside force acts on the object	

1. **Explain** How are inertia and mass related?

____

2. **Use Graphics** The hockey puck shown below is moving on the ice at a constant velocity. The arrow represents the constant velocity. In the box, draw the arrow that represents the velocity of the puck if no unbalanced forces act on the puck.

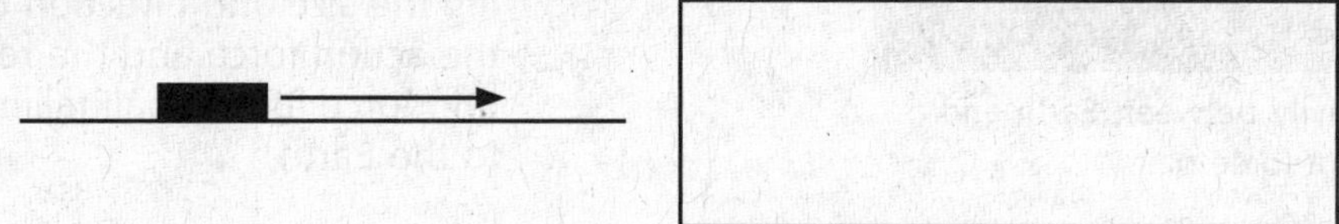

3. **Use Graphics** The ball shown below has two forces acting on it. The arrows represent the size and direction of the forces. In the box, draw the arrow that represents the net force on the ball.

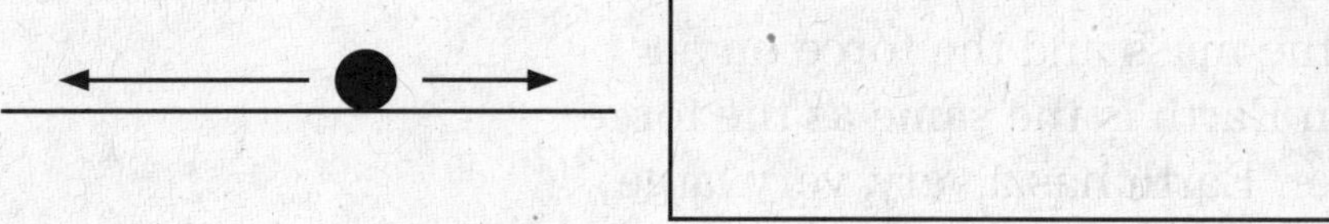

4. **Identify** Describe two things you can do to increase the acceleration of an object.

____

____

5. **Identify** Identify the action and reaction forces when you kick a soccer ball.

____

____

____

6. **Calculate** What force is needed to accelerate a 40 kg person at a rate of 4.5 m/s^2? Show your work.

____

____

____

Name _______________ Class _______________ Date _______________

CHAPTER 2 Forces and Motion

SECTION 3

Momentum

BEFORE YOU READ

After you read this section, you should be able to answer these questions:

- What is momentum?
- How is momentum calculated?
- What is the law of conservation of momentum?

What Is Momentum?

Picture a compact car and a large truck moving at the same velocity. The drivers of both vehicles put on the brakes at the same time. Which vehicle will stop first? You most likely know that it will be the car. But why? The answer is momentum.

The momentum of an object depends on the object's mass and velocity. **Momentum** is the product of the mass and velocity of an object. In the figure below, a car and a truck are shown moving at the same velocity. Because the truck has a larger mass, it has a larger momentum. A greater force will be needed to stop the truck. ☑

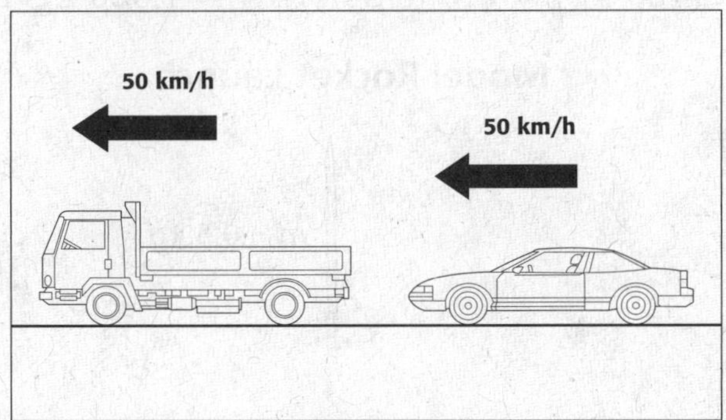

A truck and car traveling with the same velocity do not have the same momentum.

Object	Momentum
A train moving at 30 km/h	
A bird sitting on a branch high in a tree	
A truck moving at 30 km/h	
A rock sitting on a beach	

Visualize As you read, try to picture in your head the events that are described. If you have trouble imagining, draw a sketch to illustrate the event.

READING CHECK

1. Identify What is the momentum of an object?

____

____

TAKE A LOOK

2. Predict How could the momentum of the car be increased?

____

____

Critical Thinking

3. Apply Concepts Fill in the chart to the left to show which object has the most momentum, which object has a smaller amount of momentum, and which objects have no momentum.

How Can You Calculate Momentum?

If you know what an object's mass is and how fast it is going, you can calculate its momentum. The equation for momentum is

$$p = m \times v$$

In this equation, p is momentum (in kilograms multiplied by meters per second), m is the mass of the object (in kilograms), and v is the velocity of the object (in meters per second). ☑

Like velocity, momentum has direction. The direction of an object's momentum is always the same as the direction of the object's velocity.

Use the following procedure to solve momentum problems:

Step 1: Write the momentum equation.

Step 2: Replace the letters in the equation with the values from the problem.

Let's try a problem. A 120 kg ostrich is running with a velocity of 16 m/s north. What is the momentum of the ostrich?

Step 1: The equation is $p = m \times v$.

Step 2: m is 120 kg and v is 16 m/s north. So,

$$p = (120 \text{ kg}) \times (16 \text{ m/s north}) = 1{,}920 \text{ kg}\bullet\text{m/s north}$$

READING CHECK

4. Identify What do you need to know in order to calculate an object's momentum?

____

____

Math Focus

5. Calculate A 6 kg bowling ball is moving at 10 m/s down the alley toward the pins. What is the momentum of the bowling ball? Show your work.

____

Model Rocket Launch

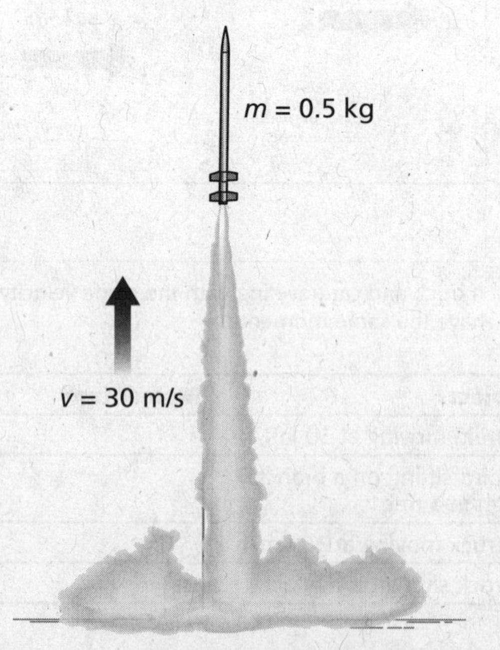

TAKE A LOOK

6. Describe Use a metric ruler to draw an arrow next to the rocket to show its momentum. The size of the rocket's momentum is 15 kg•m/s. The scale for the arrow should be 1 cm = 10 kg•m/s.

What Is the Law of Conservation of Momentum?

When a moving object hits an object at rest, some or all of the momentum of the first object is transferred to the second object. This means that the object at rest gains all or some of the moving object's momentum. During a collision, the total momentum of the two objects remains the same. Total momentum doesn't change. This is called the *law of conservation of momentum.*

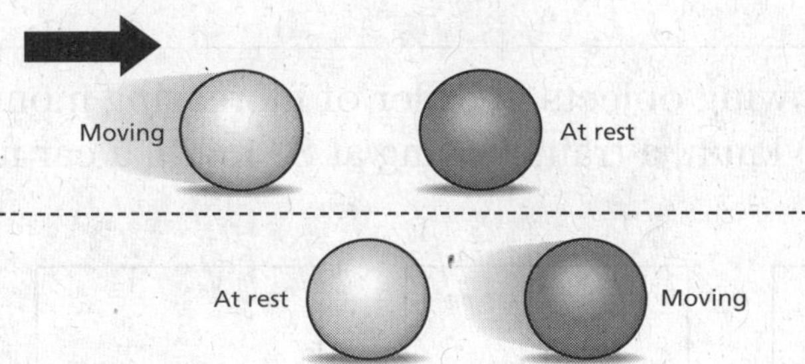

The momentum before a collision is equal to the momentum after the collision.

TAKE A LOOK

7. Identify Draw an arrow showing the size and direction of the darker ball's momentum after its collision with the lighter ball.

The law of conservation of momentum is true for any colliding objects as long as there are no outside forces. For example, if someone holds down the darker ball in the collision shown above, it will not move. In that case, the momentum of the lighter ball would be transferred to the person holding the ball. The person is exerting an outside force. ☑

The law of conservation of momentum is true for objects that either stick together or bounce off each other during a collision. In both cases, the velocities of the objects will change so that their total momentum stays the same.

READING CHECK

8. Identify When is the law of conservation of momentum not true for two objects that are interacting?

____

____

When football players tackle another player, they stick together. The velocity of each player changes after the collision because of conservation of momentum.

Although the bowling ball and bowling pins bounce off each other and move in different directions after a collision, momentum is neither gained nor lost.

Explain Words In a group, discuss how the everyday use of the word *momentum* differs from its use in science.

Name ______________________ Class ______________ Date ______________

Section 3 Review

SECTION VOCABULARY

momentum a quantity defined as the product of the mass and velocity of an object	

1. Explain A car and a train are moving at the same velocity. Do the two objects have the same momentum? Explain your answer.

____

____

____

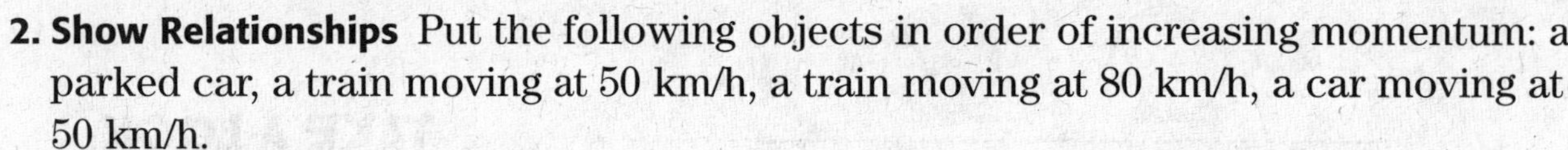

2. Show Relationships Put the following objects in order of increasing momentum: a parked car, a train moving at 50 km/h, a train moving at 80 km/h, a car moving at 50 km/h.

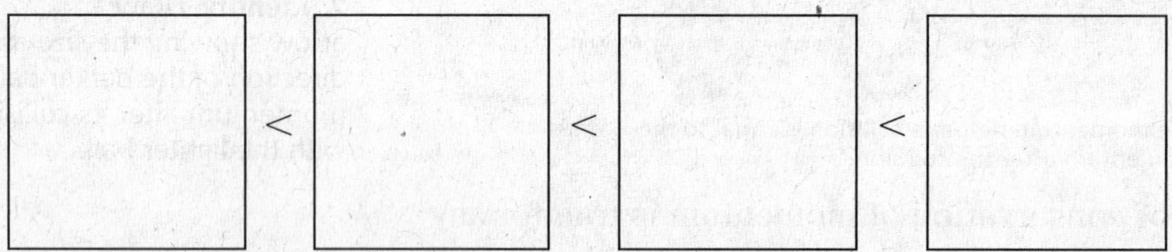

3. Calculate A 2.5 kg puppy is running with a velocity of 4.8 m/s south. What is the momentum of the puppy? Show your work.

4. Explain What is the law of conservation of momentum?

____

____

5. Calculate A ball has a momentum of 1 kg•m/s north. It hits another ball of equal mass that is at rest. If the first ball stops, what is the momentum of the other ball after the collision? (Assume there are no outside forces.) Explain your answer.

____

____

____

SECTION 1

Fluids and Pressure

BEFORE YOU READ

After you read this section, you should be able to answer these questions:

- What are fluids?
- What is atmospheric pressure?
- What is water pressure?
- What causes fluids to flow?

What Are Fluids?

You have something in common with a dog, a sea gull, and a dolphin. You and all these other animals spend a lifetime moving through fluids. A **fluid** is any material that can flow and that takes the shape of its container. Fluids have these properties because their particles can easily move past each other. Liquids and gases are fluids. ☑

Fluids produce pressure. **Pressure** is the force exerted on a given area. The motions of the particles in a fluid are what produce pressure. For example, when you pump up a bicycle tire, you push air into the tire. Air is made up of tiny particles that are always moving. When air particles bump into the inside surface of the tire, the particles produce a force on the tire. The force exerted on the area of the tire creates air pressure inside the tire.

STUDY TIP

Explain As you read this section, study each figure. In your notebook, describe what each figure tells you about pressure.

READING CHECK

1. Identify What is a fluid?

____

____

____

The air particles inside the tire hit the walls of the tire with a force. This force produces a pressure inside the tire. The pressure keeps the tire inflated.

TAKE A LOOK

2. Define What is pressure?

____

____

PRESSURE AND BUBBLES

Why are bubbles round? It's because fluids (such as the gas inside the bubbles) exert the same pressure in all directions. This gives the bubbles their round shape.

CALCULATING PRESSURE

Remember that pressure is a force exerted on an area. You can use this equation to calculate pressure:

$$pressure = \frac{force}{area}$$

The SI unit of force is the pascal. One **pascal** (Pa) is equal to a force of one newton pushing on an area of one square meter (1 N/m^2). 1 Pa of pressure is very small. A stack of 120 sheets of notebook paper exerts a pressure of about 1 Pa on a table top. Therefore, scientists usually give pressure in kilopascals (kPa). 1 kPa equals 1,000 Pa.

Let's calculate a pressure. What is the pressure produced by a book that has an area of 0.2 m^2 and a weight of 10 N? Solve pressure problems using the following procedure:

Step 1: Write the equation. $pressure = \frac{force}{area}$

Step 2: Substitute and solve. $= \frac{10\ N}{0.2\ m^2} = 50\ \frac{N}{m^2} = 50\ Pa$

Math Focus

3. Calculate What pressure is exerted by a crate with a weight of 3,000 N on an area of 2 m^2? Show your work.

What Is Atmospheric Pressure?

The *atmosphere* is the layer of gases that surrounds Earth. Gravity holds the atmosphere in place. The pull of gravity gives air weight. The pressure caused by the weight of the atmosphere is called **atmospheric pressure**.

Atmospheric pressure is exerted on everything on Earth, including you. At sea level, the pressure is about 101,300 Pa (101.3 kPa). This means that every square centimeter of your body has about 10 N (2 lbs) of force pushing on it.

Why doesn't your body collapse under this pressure? Like the air in a balloon, the fluids inside your body exert pressure. This pressure inside your body acts against the atmospheric pressure.

TAKE A LOOK

4. Describe What would be the length of the arrows if the balloon were inflated more? Explain your answer.

____

____

____

____

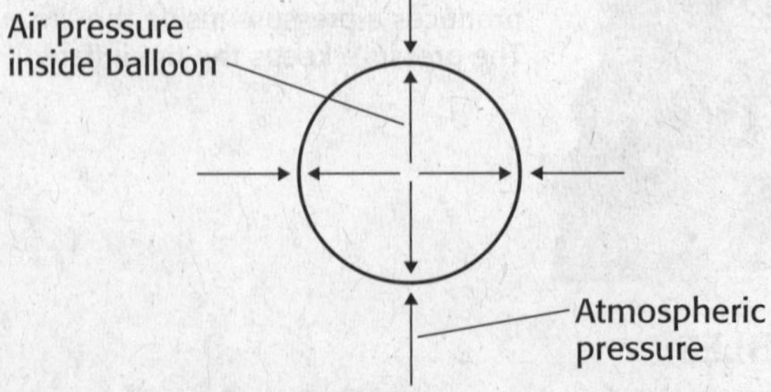

Air pressure inside balloon

Atmospheric pressure

The air inside the balloon produces a pressure inside the balloon. The pressure inside the balloon equals the atmospheric pressure outside the balloon. Therefore, the balloon stays inflated.

PRESSURE, ALTITUDE, AND DEPTH

It is very difficult to climb Mount Everest. One reason is that there is not very much air at the top of Mount Everest. The atmospheric pressure on top of Mount Everest is only about one-third of that at sea level. As you climb higher, the pressure gets lower and lower. At the top of the atmosphere, the pressure is almost 0 Pa. ☑

READING CHECK

5. Describe As altitude increases, what happens to atmospheric pressure?

____

Increasing altitude

12,000 m above sea level	Airplanes fly at about 12,000 m above sea level. Atmospheric pressure there is about 20 kPa.
9,000 m above sea level	The top of Mount Everest is about 9,000 m above sea level. Atmospheric pressure there is about 30 kPa.
4,000 m above sea level	La Paz, the capital of Bolivia, is about 4,000 m above sea level. Atmospheric pressure in La Paz is about 51 kPa.
0 m above sea level	At sea level, atmospheric pressure is about 101 kPa.

Math Focus

6. Calculate About what fraction of atmospheric pressure at sea level is atmospheric pressure at La Paz?

____

Air pressure is greatest at Earth's surface because the entire weight of the atmosphere is pushing down there. This is true for all fluids. As you get deeper in a fluid, the pressure gets higher. You can think of being at sea level as being "deep" in the atmosphere. ☑

READING CHECK

7. Explain Why is atmospheric pressure greatest at the surface of Earth?

____

____

____

PRESSURE CHANGES AND YOUR BODY

What happens to your body when atmospheric pressure changes? You may have felt your ears "popping" when you were in an airplane or in a car climbing a mountain. Air chambers behind your ears help to keep the pressure in your ears equal to air pressure. The "pop" happens because the pressure inside your ears changes as air pressure changes.

What Affects Water Pressure?

Water is a fluid. Therefore, it exerts a pressure. Like air pressure, water pressure increases as depth increases, as shown in the figure below. The pressure increases as the diver gets deeper because more and more water is pushing on her. In addition, the atmosphere pushes down on the water. Therefore, the total pressure on the diver is the sum of the water pressure and the atmospheric pressure. ☑

READING CHECK

8. Explain Why does pressure increase as depth increases?

____

____

____

Discuss In a small group, talk about the kinds of adaptations that deep-water organisms, such as the viper fish, may have to help them survive at very high water pressures.

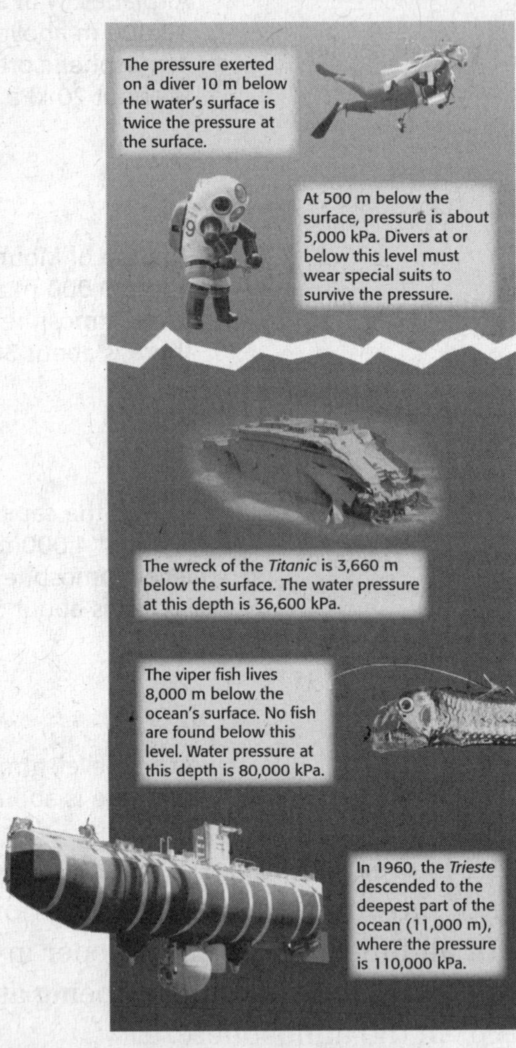

DENSITY EFFECTS ON WATER PRESSURE

Density is a measure of how closely packed the particles in a substance are. It is a ratio of the mass of an object to its volume. Water is about 1,000 times denser than air. Water has more mass (and weighs more) than the same volume of air. Therefore, water exerts more pressure than air. The pressure exerted by 10 m of water is 100 kPa. This is almost the same as the pressure exerted by the whole atmosphere.

Critical Thinking

9. Infer What is the total pressure in kPA 10 m below the water? Hint: the total pressure is the sum of the atmospheric pressure and the water pressure.

____

What Causes Fluids to Flow?

All fluids flow from areas of high pressure to areas of low pressure. Imagine a straw in a glass of water. Before you suck on the straw, the air pressure inside the straw is equal to the air pressure on the water. When you suck on the straw, the air pressure inside the straw decreases. However, the pressure on the water outside the straw stays the same. The pressure difference forces water up the straw and into your mouth.

Critical Thinking

10. Apply Concepts Why does the air pressure inside a straw go down when you suck on the straw?

____

____

____

____

____

PRESSURE DIFFERENCE AND BREATHING

The flow of air from high pressure to low pressure is also what allows you to breathe. In order to inhale, a muscle in your chest moves down. This makes the volume of your chest bigger, so your lungs have more room to expand. As your lungs expand, the pressure inside them goes down. Atmospheric pressure is now higher than the pressure inside your lungs, so air flows into your lungs. The reverse of this process happens when you exhale, as shown in the figure below.

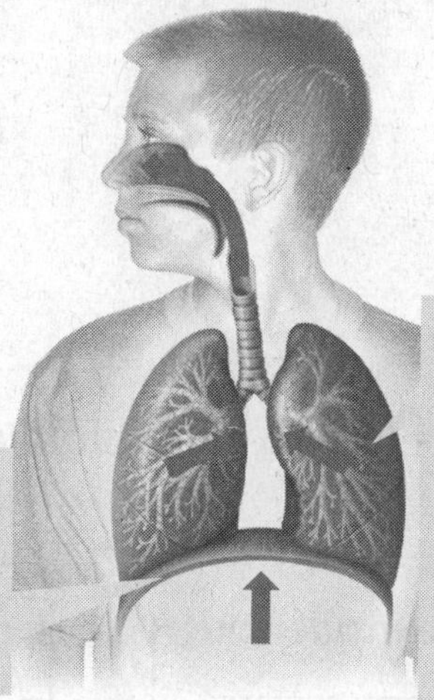

When you exhale, a muscle in your chest moves upward. The volume of your chest decreases.

As the volume of your chest decreases, the pressure in your lungs increases. The pressure in your lungs becomes greater than the pressure outside your lungs. Therefore, the air flows out of your lungs (higher pressure) into the air (lower pressure).

TAKE A LOOK

11. Explain Why does air flow out of your lungs when you exhale?

____

____

____

PRESSURE DIFFERENCES AND TORNADOES

During a tornado, wind speeds can reach 300 miles per hour or more! Some of the damaging winds caused by a tornado are due to pressure differences. The air pressure inside a tornado is very low. Because the air pressure outside the tornado is high, the air rushes into the tornado and produces strong winds. The winds cause the tornado to act as a giant vacuum cleaner. Objects are pulled in and lifted up by these winds.

Name ______________________ Class ______________ Date ______________

Section 1 Review

SECTION VOCABULARY

atmospheric pressure the pressure caused by the weight of the atmosphere **fluid** a nonsolid state of matter in which the atoms or molecules are free to move past each other, as in a gas or liquid	**pascal** the SI unit of pressure (symbol, Pa) **pressure** the amount of force exerted per unit area of a surface

1. Describe How do fluids exert pressure on a container?

____

____

2. Evaluate Define density in terms of mass and volume. How does density affect pressure?

____

____

____

3. Calculate The water in a glass has a weight of 2.5 N. The bottom of the glass has an area of 0.012 m^2. What is the pressure exerted by the water on the bottom of the glass? Show your work.

4. Describe Fill in the blank spaces in the chart below to show how air moves in and out of your lungs when you breathe.

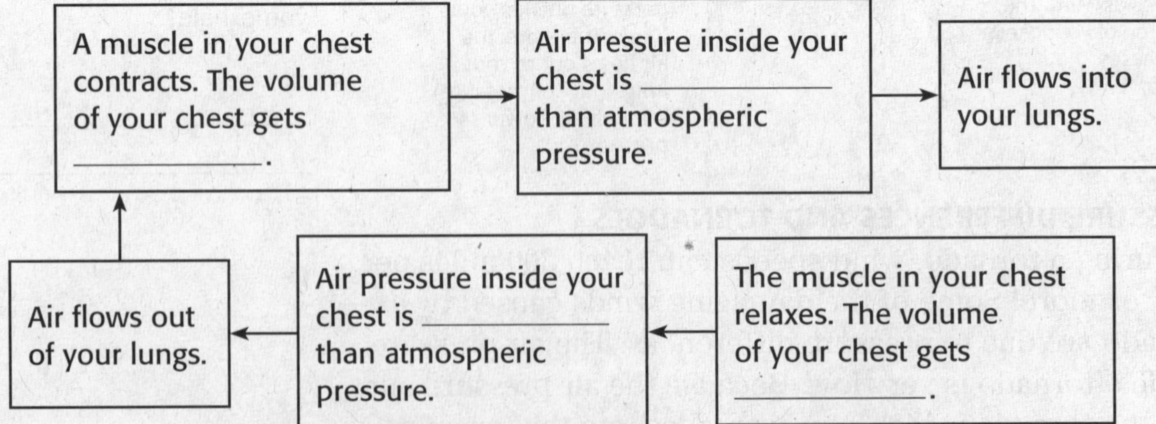

Name ______________________ Class ______________ Date ____________

CHAPTER 3 Forces in Fluids

SECTION 2

Buoyant Force

BEFORE YOU READ

After you read this section, you should be able to answer these questions:

- What is buoyant force?
- What makes objects sink or float?
- How can we change an object's density?

National Science Education Standards
PS 1a, 2c

What Is Buoyant Force and Fluid Pressure?

Why does an ice cube that has been pushed under the water pop back up? A force called buoyant force pushes the ice cube up to the water's surface. **Buoyant force** is the upward force that a fluid exerts on all objects in the fluid. If an object is buoyant, that means it will float on water like a raft. Or rise in the air like a helium-filled balloon.

Look at the figure below. Water exerts a pressure on all sides of the object in the water. The water produces the same amount of horizontal force on both sides of the object. These equal forces balance one another.

However, the vertical forces are not equal. Remember that fluid pressure increases with depth. There is more pressure on the bottom of the object than on the top. ☑

The longer arrows in the figure below show the larger pressures. You can see that the arrows are longest underneath the object. This shows that the water applies a net upward force on the object. This upward force is buoyant force. It is what makes the object float.

STUDY TIP

Learn New Words As you read, underline words you don't understand. When you figure out what they mean, write the words and their definitions in your notebook.

READING CHECK

1. Identify Why is the force on the bottom of an object in a fluid larger than the force on the top?

____

____

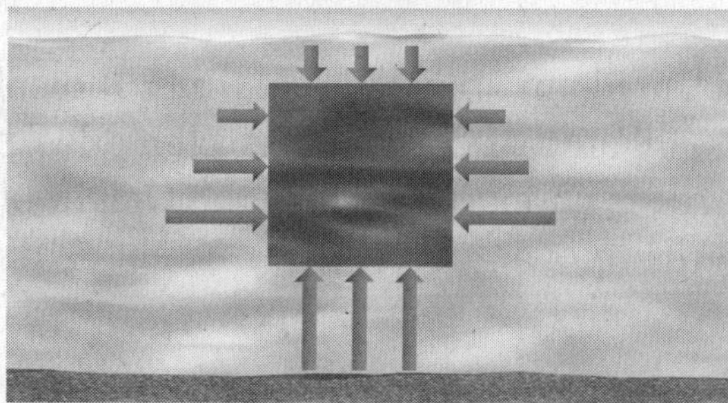

There is more pressure at the bottom of an object because pressure increases with depth. The differences in pressure produce an upward buoyant force on the object.

TAKE A LOOK

2. Identify What produces buoyant force?

____

____

____

____

Buoyant force is what makes you feel lighter when you float in a pool of water. The buoyant force of the water pushes up on your body and reduces your weight.

DETERMINING BUOYANT FORCE

Archimedes, a Greek mathematician who lived in the third century BCE, discovered how to find buoyant force. Archimedes found that objects in water *displace*, or take the place of, water. The weight of the displaced water equals the buoyant force of the water. This is now known as **Archimedes' principle**.

You can find buoyant force by measuring the weight of the water that an object displaces. Suppose a block of ice displaces 250 mL of water. The weight of 250 mL of water is about 2.5 N. The weight of the displaced water equals the buoyant force. Therefore, the buoyant force on the block is 2.5 N.

Notice that only the weight of the displaced fluid determines the buoyant force on an object. The weight of the object does not affect buoyant force.

Math Focus

3. Calculate A can of soda displaces about 360 mL of water when it is put in a tank of water. The weight of 360 mL of water is about 3.6 N. What is the buoyant force on the can of soda?

____

What Makes Objects Float or Sink?

An object in a fluid will sink if its weight is greater than the buoyant force. An object floats only when the buoyant force is equal to or less than the object's weight. ☑

READING CHECK

4. Identify Will an object sink or float if its weight is less than the buoyant force?

____

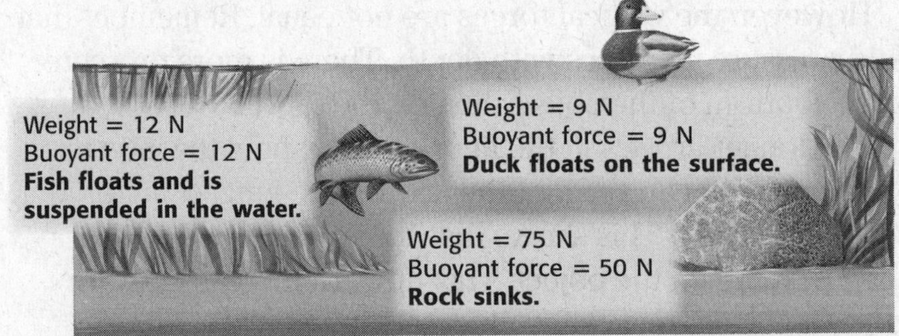

TAKE A LOOK

5. Explain Why does the fish float in the middle of the water?

____

____

____

SINKING

The rock in the figure above weighs 75 N. It displaces 5 L, or about 50 N, of water. According to Archimedes' principle, the buoyant force is about 50 N. Since the weight of the rock is greater than the buoyant force, the rock sinks.

FLOATING

The fish in the figure weighs 12 N. It displaces a volume of water that weighs 12 N. The buoyant force and the fish's weight are equal, so the fish floats in the water. It does not sink to the bottom or rise to the surface—it is *suspended* in the water.

Critical Thinking

6. Apply Concepts If the duck in the figure weighed 10 N, would more or less of the duck be underwater?

____

BUOYING UP

If the duck dove underwater, it would displace more than 9 N of water. As a result, the buoyant force on the duck would be greater than the duck's weight. When the buoyant force on an object is greater than the object's weight, the object is *buoyed up*, or pushed up in the water.

An object is buoyed up until the part underwater displaces an amount of water that equals the object's weight. Therefore, the part of the duck that is underwater displaces 9 N of water.

STANDARDS CHECK

PS 2c If more than one force acts on an object along a straight line, then the forces will reinforce or cancel one another, depending on their direction and magnitude. Unbalanced forces will cause changes in the speed or direction of an object's motion.

7. Explain If a log floating on the water is pushed underwater, why will it pop back up?

____

____

____

How Does Density Affect Floating?

Remember that *density* is the mass of an object divided by its volume. How does the density of the rock compare to the density of water? The volume of the rock is 5 L, and it displaces 5 L of water. The weight of the rock is 75 N, and the weight of 5 L of water is 50 N. The weight of an object is a measure of its mass. In the same volume, the rock has more mass than water. Therefore, the rock is more dense than water. ☑

The rock sinks because it is denser than water. The duck floats because it is less dense than water. The fish floats suspended in the water because it has the same density as the water. ☑

READING CHECK

8. Define What is density?

____

____

READING CHECK

9. Explain How does an object's density determine whether it floats or sinks?

____

____

____

MORE DENSE OR LESS DENSE THAN AIR

Why does an ice cube float on water but not in air? An ice cube floats in water because it is less dense than water. However, most substances are more dense than air. The ice cube is more dense than air, so it does not float in air.

One substance that is less dense than air is helium, a gas. When a balloon is filled with helium, the filled balloon becomes less dense than air. Therefore, the balloon floats in air, like the one in the picture below.

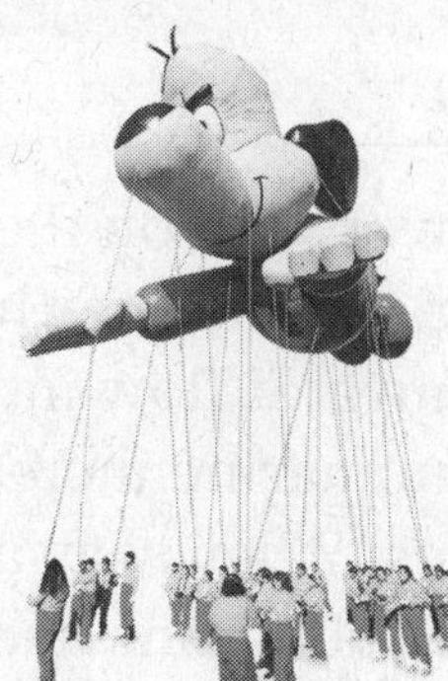

This balloon floats because the helium in it is less dense than air.

TAKE A LOOK

10. Infer The balloon in the picture is filled with about 420,000 L of helium. Does 420,000 L of air have a greater or smaller mass than the 420,000 L of helium in the balloon? Explain your answer.

____

____

____

____

What Affects an Object's Density?

The total density of an object can change if its mass or volume changes. If volume increases and mass stays the same, density decreases. If mass increases and volume stays the same, density increases.

A block of steel is denser than water, so it sinks. If that block is shaped into a hollow form, the overall density of the form is less than water. Therefore, the ship floats.

TAKE A LOOK

11. Compare What is the volume of the ship compared to the volume of the steel used to make the ship?

____

____

____

____

CHANGING SHAPE

Steel is almost eight times denser than water. Yet huge steel ships cruise the oceans with ease. If steel is more dense than water, how can these ships float? The reason a steel ship floats has to do with its shape. If the ship were just a big block of steel, it would sink very quickly. However, ships are built with a hollow shape. The hollow shape increases the volume that the steel takes up without increasing the mass of the steel.

Increasing the volume of the steel produces a decrease in its density. When the volume of the ship becomes large enough, the overall density of the ship becomes less than water. Therefore, the ship floats. ☑

Most ships are built to displace more water than is necessary for the ship to float. These ships are made this way so that they won't sink when people and cargo are loaded onto the ship.

READING CHECK

12. Explain How can changing the shape of an object lower its overall density?

____

____

____

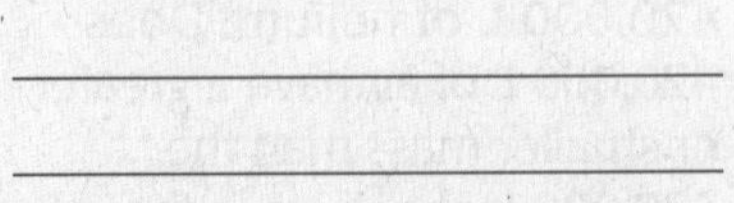

CHANGING MASS

A submarine is a ship that can travel both on the surface of the water and underwater. Submarines have *ballast tanks* that can open to let seawater flow in. When seawater flows in, the mass of the submarine increases. Therefore, its overall density increases. When seawater is pushed out, the overall density of the submarine decreases and it rises to the surface.

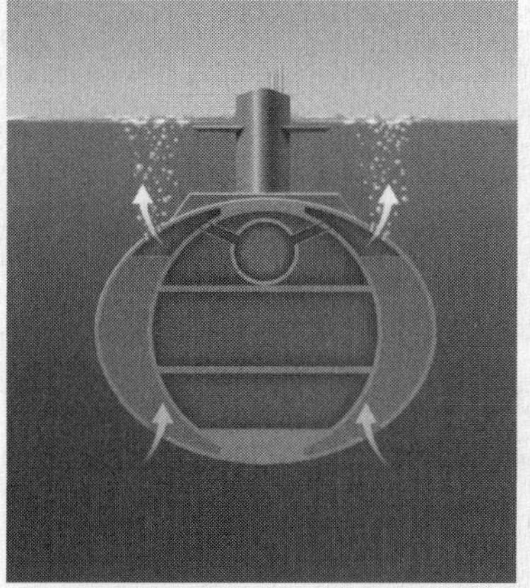

Water flows into the ballast tanks. The submarine becomes more dense and sinks.

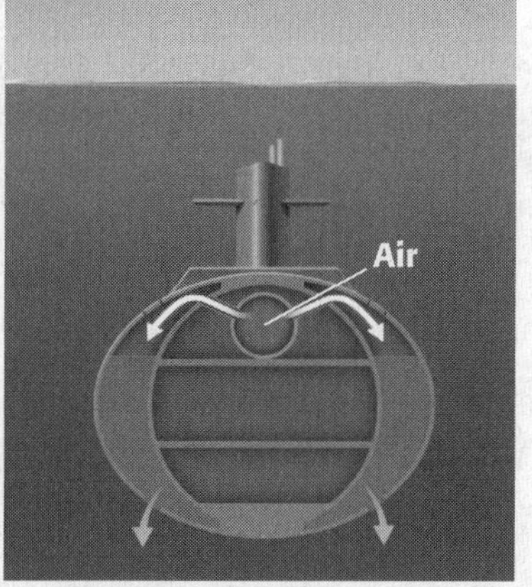

Compressed air forces water out of the ballast tanks. The submarine becomes less dense and floats to the surface.

TAKE A LOOK

13. Describe How does a submarine increase its density?

____

____

CHANGING VOLUME

Some fish can change their overall density by changing their volume. Most bony fish have an organ called a *swim bladder*. This swim bladder can fill with gases or release gases. The gases are less dense than the rest of the fish. When gases go into the swim bladder, the overall volume of the fish increases, but the mass of the fish does not change as much. This lowers the overall density of the fish and keeps it from sinking in the water. ☑

The fish's nervous system controls the amount of gas in the bladder. Some fish, such as sharks, do not have a swim bladder. These fish must swim constantly to keep from sinking.

14. Explain How do most bony fish change their overall density?

____

____

____

Most bony fish have a swim bladder, an organ that allows them to adjust their overall density.

Name ________________ Class ________________ Date ________________

Section 2 Review

NSES PS 1a, 2c

SECTION VOCABULARY

Archimedes' Principle the principle that states that the buoyant force on an object in a fluid is an upward force equal to the weight of the volume of fluid that the object displaces	**buoyant force** the upward force that keeps an object immersed in or floating on a liquid

1. Predict In Figure 1, a block of wood is floating on the surface of some water. In Figure 2, the same block of wood is pushed beneath the surface of the water. In the space below, predict what will happen to the wood when the downward force in Figure 2 is removed. Use the term *buoyant force* in your answer.

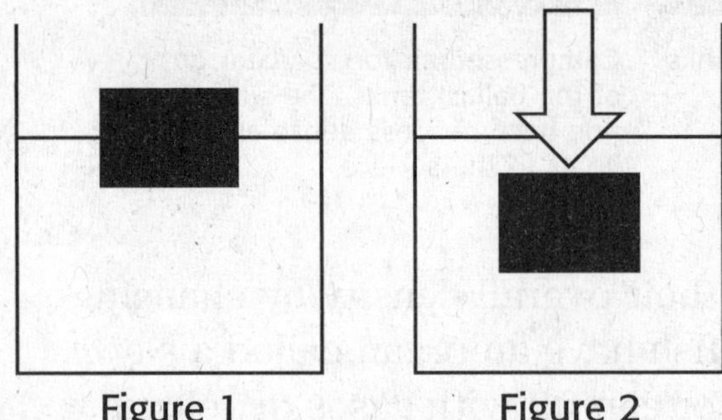

____

____

____

2. Calculate A container that is filled with mercury has a mass of 4810 g. If the volume of the container is 355 mL, what is its overall density? Show your work. Round your answer to the nearest tenth.

3. Identify Give two ways that an object's overall density can change.

____

____

4. Explain How can knowing an object's density help you to predict whether the object will float or sink in a fluid?

____

____

____

Name ______________________ Class ______________ Date ____________

CHAPTER 3 Forces in Fluids

SECTION 3

Fluids and Motion

BEFORE YOU READ

After you read this section, you should be able to answer these questions:

- How does fluid speed affect pressure?
- How do lift, thrust, and wing size affect flight?
- What is drag?
- What is Pascal's principle?

National Science Education Standards

PS 1a

What Are Fluid Speed and Pressure?

Usually, when you think of something flowing, it is a liquid such as water. But remember that a fluid is any material that can flow and that takes the shape of its container. So, gases such as air are fluids, too. Both liquids and gases flow when forces act on them.

An 18th century Swiss mathematician Daniel Bernoulli, found that fast-moving fluids have a lower pressure than slow-moving fluids. **Bernoulli's principle** states that as the speed of a moving fluid increases, the fluid's pressure decreases. ☑

Have you ever watched an airplane take off and wondered how it could stay in the air? Look at the picture of the airplane below and you'll see Bernoulli's principle at work.

STUDY TIP

Reading Organizer As you read this section, create an outline of the section. Use headings from the section in your outline.

READING CHECK

1. Describe What is Bernoulli's principle?

____

____

____

____

Wing Design and Lift

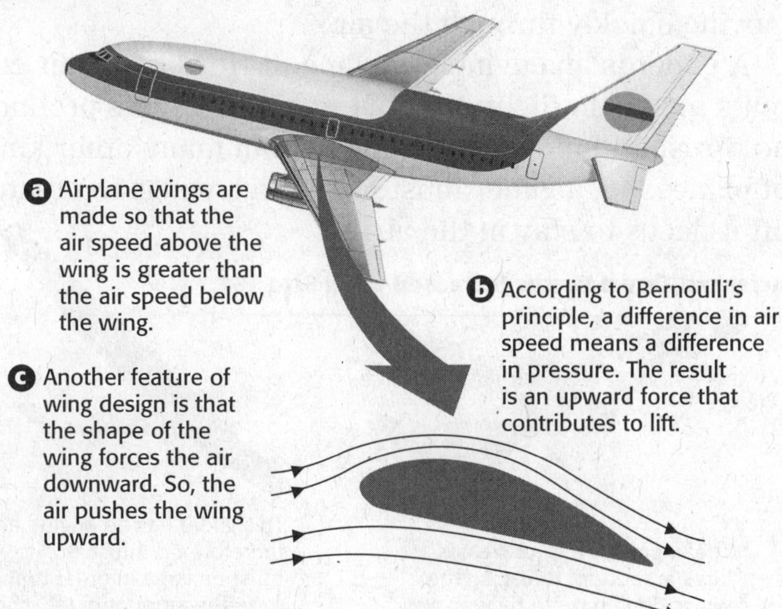

Critical Thinking

2. Infer Why do you need a windy day to fly a kite?

____

____

____

____

What Factors Affect Flight?

A common airplane in the skies today is the Boeing 737 jet. Even without passengers, the plane weighs 350,000 N (about 79,000 lbs). How can something so big and heavy get off the ground? Wing shape plays a role in helping these big planes, as well as smaller planes, fly.

According to Bernoulli's principle, the fast-moving air above the wing exerts less pressure than the slow-moving air below the wing. As a result, the greater pressure below the wing exerts an upward force. This upward force, known as **lift**, pushes the wings (and the rest of the airplane or bird) upward against the downward pull of gravity. ☑

3. Explain What is lift?

______

______

______

THRUST AND LIFT

The amount of lift caused by a plane's wing is determined partly by the speed of the air around the wing. Thrust determines the speed of a plane. **Thrust** is the forward force that the plane's engine produces. Usually, a plane with a large amount of thrust moves faster than a plane that has less thrust. This faster speed means greater lift.

WING SIZE, SPEED, AND LIFT

The size of a plane's wings also affects the plane's lift. The jet plane in the picture below has small wings, but its engine gives a large amount of thrust. This thrust pushes the plane through the sky at great speeds. As a result, the jet creates a large amount of lift with small wings by moving quickly through the air.

A glider is an engineless plane. It rides rising air currents to stay in flight. Without engines, gliders produce no thrust and move more slowly than many other kinds of planes. So, a glider must have large wings to create the lift it needs to stay in the air. ☑

4. Identify What are two factors that affect lift?

______

______

Increased Thrust Versus Increased Wing Size

The engine of this jet creates a large amount of thrust, so the wings don't have to be very big.

This glider has no engine and therefore no thrust. So, its wings must be large in order to maximize the amount of lift achieved.

BERNOULLI, BIRDS, AND BASEBALL

Birds don't have engines, so they must flap their wings in order to supply thrust and lift. A small bird must flap its wings at a fast pace to stay in the air. But a hawk flaps its wings only occasionally. It flies with little effort because it has larger wings. Fully extended, a hawk's wings allow it to glide on wind currents and still have enough lift to stay in the air. ☑

5. Identify What does a bird supply by flapping its wings?

____

Bernoulli's Principle and the Screwball

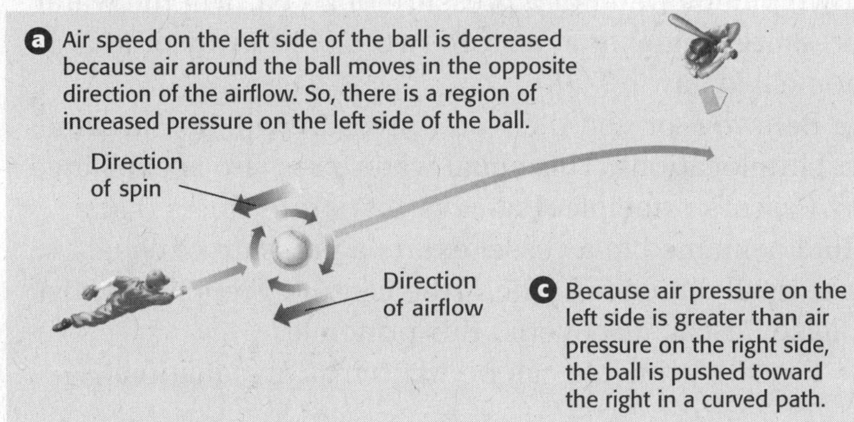

TAKE A LOOK

6. Identify Suppose the pitcher threw the ball and it curved toward the batter. Would the ball be spinning in a clockwise or counterclockwise direction?

____

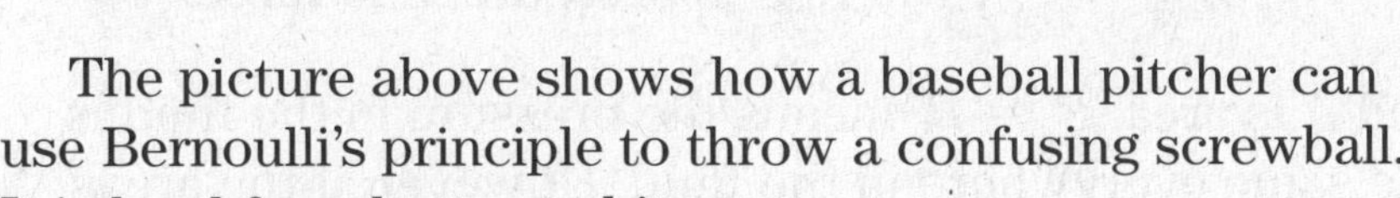

The picture above shows how a baseball pitcher can use Bernoulli's principle to throw a confusing screwball. It is hard for a batter to hit.

DRAG AND MOTION IN FLUIDS

Have you ever walked into a strong wind and noticed that the wind seemed to slow you down or push you backward? Fluids exert a force that opposes the motion of objects moving through them. The force in a fluid that opposes or restricts motion is called **drag**. ☑

In a strong wind, air "drags" on your body and makes it difficult for you to move into the wind. Drag also works against the forward motion of a plane or bird in flight. Drag is usually caused by an irregular flow of air. An irregular or unpredictable flow of fluids is known as *turbulence*.

READING CHECK

7. Decribe What is drag?

____

____

____

TURBULENCE AND LIFT

Turbulence causes drag and that reduces lift. Drag can be a serious problem for airplanes moving at high speeds. As a result, airplanes have ways to reduce turbulence. For example, airplanes have flaps on their wings. When the flaps move, it changes the shape or area of a wing. This change can reduce drag and increase lift. ☑

READING CHECK

8. Explain How do airplanes reduce turbulence?

____

____

____

What Is Pascal's Principle?

Imagine that the water-pumping station in your town pumps water at a pressure of 20 Pa. Will the water pressure be higher at a store two blocks away or at a home 2 km away?

Believe it or not, the water pressure will be the same at both locations. This equal water pressure is explained by Pascal's principle. **Pascal's principle** states that a fluid contained in a vessel exerts a pressure of equal size in all directions. The 17the century French scientist, Blaise Pascal, discovered this principle.

Critical Thinking

9. Infer Would Pascal's principle still apply if there is a leak in the town's water system? Explain your answer.

____

____

____

____

Pascal's principle can be written as an equation:

$P_1 = P_2$ or $\frac{F_1}{A_1} = \frac{F_2}{A_2}$, where P is pressure, F is force, and A is area. $P_1 = P_2$ means the pressure in the fluid is the same everywhere in the fluid. However, if the areas pushed on by a fluid are different, the forces will be different. In the figure below, Area 2 is larger than Area 1.

Math Focus

10. Determine If force 1 = 20 N, area 1 = 5 cm², and area 2 = 200 cm², what is force 2? Show your work.

Suppose that force 1 = 10 N, area 1 = 2 cm², and area 2 = 100 cm², what is force 2? Rearrange the equation and put in the values.

$$F_2 = A_2 \times \frac{F_1}{A_1} = 100\text{ cm}^2 \times \frac{10\text{ N}}{2\text{ cm}^2} = 500\text{ N}$$

Pushing on one end of the fluid with a 10 N force caused a 500 N force on the other end. This will be used on the next page to explain how car brakes work.

PASCAL'S PRINCIPLE AND MOTION

Hydraulic devices use Pascal's principle to move or lift objects. Hydraulic means the devices operate using fluids, usually oil. In hydraulic devices liquids cannot be easily compressed, or squeezed, into a smaller space. Cranes, forklifts, and bulldozers have hydraulic devices that help them lift heavy objects.

Hydraulic machines can multiply forces. Car brakes are a good example of this. In the picture below, a driver's foot exerts pressure on a cylinder of liquid. This pressure is transmitted to all parts of the liquid-filled brake system. The liquid moves the brake pads. The pads press against the wheels and friction stops the car. ☑

The force is multiplied. This is because the pistons that push the brake pads are larger than the piston pushed by the brake pedal.

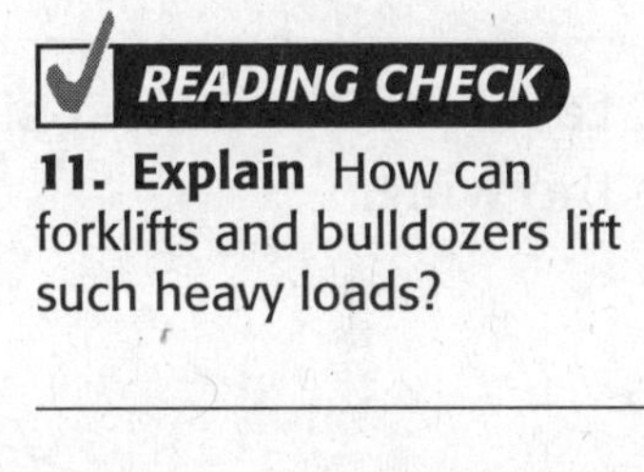

11. Explain How can forklifts and bulldozers lift such heavy loads?

____

____

____

Because of Pascal's principle, the touch of a foot can stop tons of moving metal.

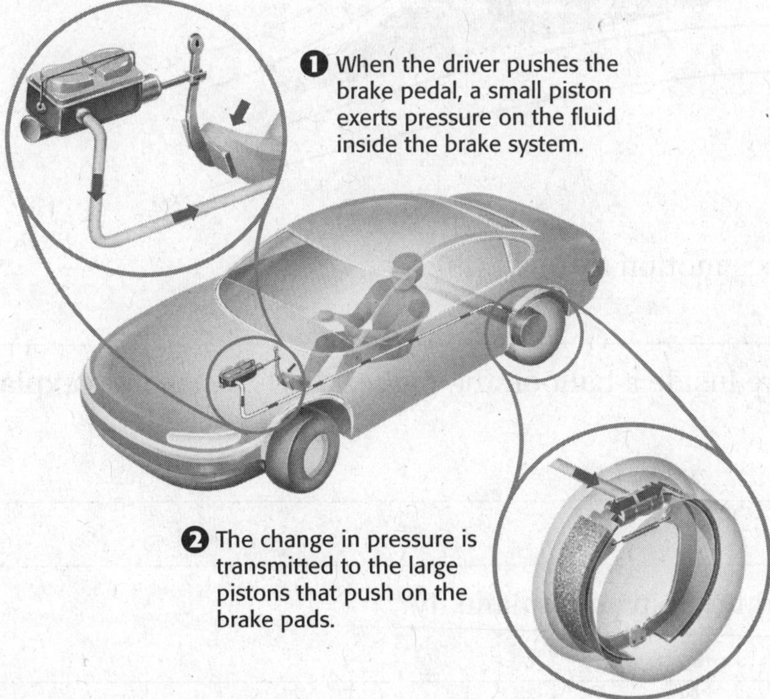

Name ______________________ Class ______________ Date ______________

Section 3 Review

NSES PS 1a

SECTION VOCABULARY

Bernoulli's principle the principle that states that the pressure in a fluid decreases as the fluid's velocity increases

drag a force parallel to the velocity of the flow; it opposes the direction of an aircraft and, in combination with thrust, determines the speed of the aircraft

lift a upward force on an object that moves in a fluid

Pascal's principle the principle that states that a fluid in equilibrium contained in a vessel exerts a pressure of equal intensity in all direction

thrust the pushing or pulling force exerted by the engine of an aircraft or rocket

1. Explain What is the relationship between pressure and fluid speed?

____

____

2. Label Write "fastest air speed" and "highest pressure" in the correct places on the wing.

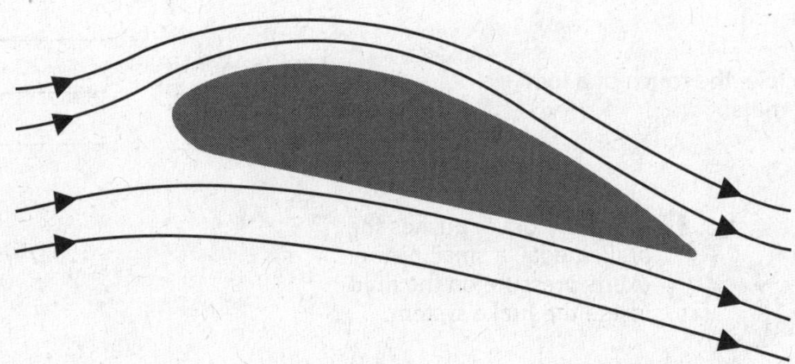

3. Identify What force opposes motion through fluid?

____

4. Infer Where is the pressure inside a balloon the highest? What principle explains your answer?

____

____

5. Explain How do thrust and lift help an airplane fly?

____

____

____

Name ______________________ Class ______________ Date __________

CHAPTER 4 Work and Machines

SECTION 1 Work and Power

BEFORE YOU READ

After you read this section, you should be able to answer these questions:

- What is work?
- How do we measure work?
- What is power and how is it calculated?

National Science Education Standards
PS 3a

What Is Work?

You may think of work as a large homework assignment. You have to read a whole chapter by tomorrow. That sounds like a lot of work, but in science, work has a different meaning. **Work** is done when a force causes an object to move in the direction of the force. You might have to do a lot of thinking, but you are not using a force to move anything. You are doing work when you to turn the pages of the book or move your pen when writing. ☑

As you read this section, write the questions in Before You Read in your science notebook and answer each one.

The student in the figure below is bowling. She is doing work. She applies a force to the bowling ball and the ball moves. When she lets go of the ball she stops doing work. The ball keeps rolling but she is not putting any more force on the ball.

READING CHECK

1. Describe When is work done on an object?

____

____

____

____

You might be surprised to find out that bowling is work!

The direction of the force

The direction of the bowling ball

How Is Energy Transferred When Work Is Done?

The bowler in the figure above has done work on the bowling ball. Since the ball is moving, it now has *kinetic* energy. The bowler has transferred energy to the ball.

STANDARDS CHECK

PS 3a Energy is a property of many substances and is associated with heat, light, electricity, mechanical energy, motion, sound, nuclei, and the nature of a chemical. Energy is transferred in many ways.

2. Identify How is energy transmitted to a bowling ball to make it move down the alley?

____

____

When Is Work Done On An Object?

Applying a force does not always mean that work is done. If you push a car but the car does not move, no work is done on the car. Pushing the car may have made you tired. If the car has not moved, no work is done on the car. When the car moves, work is done. If you apply a force to an object and it moves in that direction, then work is done.

How Are Work and Force Different?

You can apply a force to an object, but not do work on the object. Suppose you are carrying a heavy suitcase through an airport. The direction of the force you apply to hold the suitcase is up. The suitcase moves in the direction you are walking. The direction the suitcase is moving is not the same as the direction of the applied force. So when you carry the suitcase, no work is done on the suitcase. Work is done when you lift the suitcase off the ground.

Work is done on an object if two things happen.

1. An object moves when a force is applied.
2. The object moves in the direction of the force. ☑

In the figure below, you can see how a force can cause work to be done.

READING CHECK

3. Describe What two things must happen to do work on an object?

____

____

____

____

____

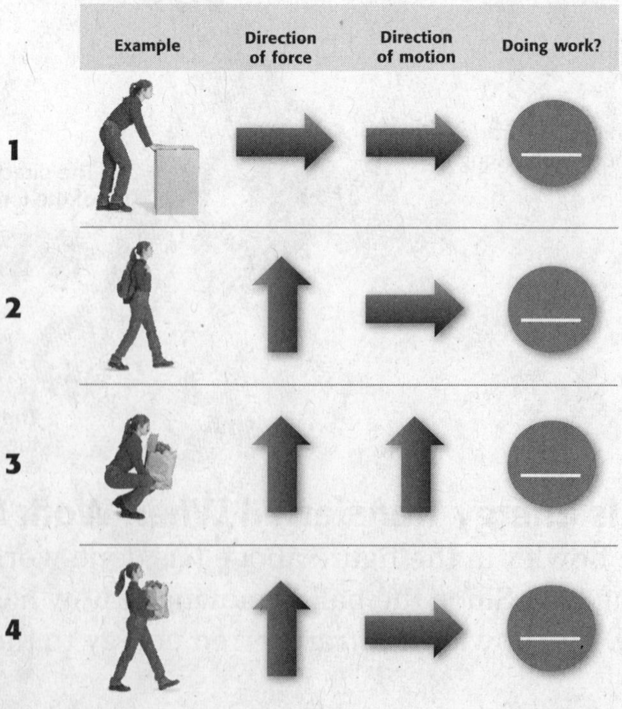

	Example	Direction of force	Direction of motion	Doing work?
1				
2				
3				
4				

TAKE A LOOK

4. Identify In the figure, there are four examples. For each example, decide if work is being done and write your answers in the figure.

How Is Work Calculated?

An equation can be written to calculate the work (W) it takes to move an object. The equation shows how work, force, and distance are related to each other:

$$W = F \times d$$

In the equation, F is the force applied to an object (in newtons). d is the distance the object moves in the direction of the force (in meters). The unit of work is the newton-meter (N×m). This is also called a **joule** (J). When work is done on an object, energy is transferred to the object. The joule is a unit of energy.

Let's try a problem. How much work is done if you push a chair that weighs 60 N across a room for 5 m?

Step 1: Write the equation.

$$W = F \times d$$

Step 2: Place values into the equation, and solve for the answer.

$$W = 60 \text{ N} \times 5 \text{ m} = 300 \text{ J}$$

The work done on the chair is 300 joules.

Math Focus

5. Calculate How much work is done pushing a car 20 m with a force of 300 N? Show your work.

80 N

1 meter

***W* = 80 N × 1 m = 80 J**

The force to lift an object is the same as the force of gravity on the object. In other words, the object's weight is the force.

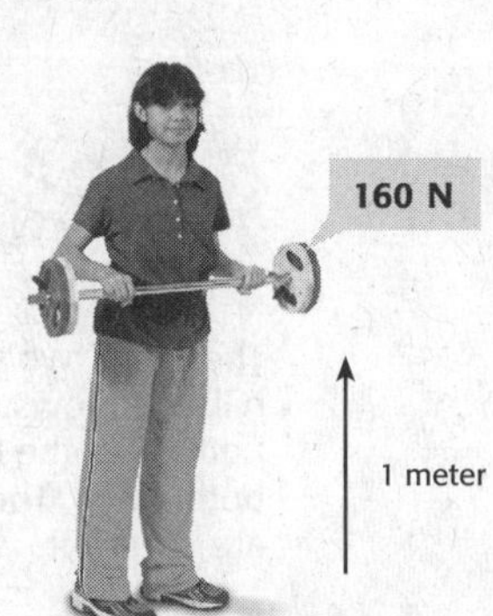

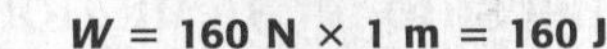

***W* = 160 N × 1 m = 160 J**

The amount of work increases when the weight of an object increases. More force is needed to lift the object.

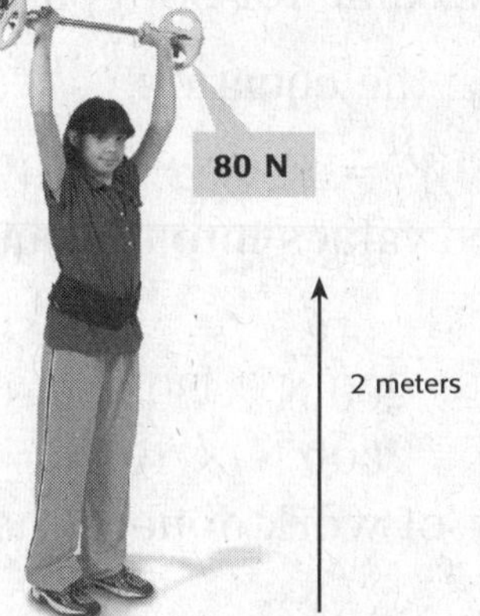

***W* =** ______________________

The amount of work also increases when the distance increases.

Math Focus

6. Calculate In the last figure, a barbell weighing 80 N is lifted 2 m off of the ground. How much work is done? Show your work.

Two Paths, Same Work?

A car is pushed to the top of a hill using two different paths. The first path is a long road that has a low, gradual slope. The second path is a steep cliff. Pushing the car up the long road doesn't need as much force as pulling it up the steep cliff. But, believe it or not, the same amount of work is done either way. It is clear that you would need a different amount of force for each path.

Look at the figure below. Pushing the car up the long road uses a smaller force over a larger distance. Pulling it up the steep cliff uses a larger force over a smaller distance. When work is calculated for both paths, you get the same amount of work for each path. ☑

READING CHECK

7. Explain Suppose it takes the same amount of work for two people to move an object. One person applies less force in moving the object than the other. How can they both do the same amount of work?

____

____

____

____

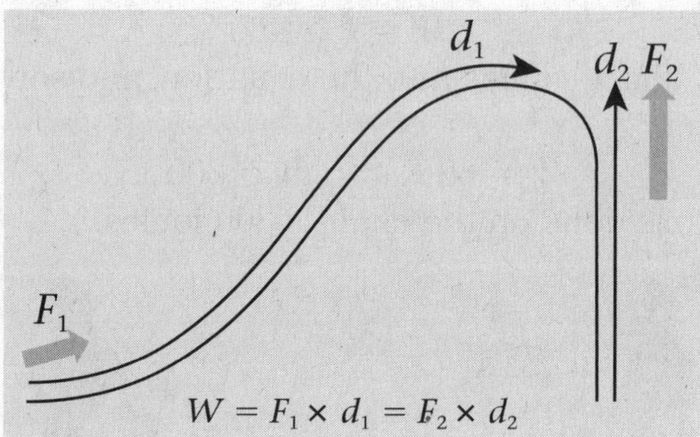

There are two paths to move the car to the top of the hill. d_1 follows the shape of the hill. d_2 is straight from the bottom to the top of the hill. The same work is done in both paths. The distance and force for each of the paths are different.

Let's do a calculation. Suppose path 1 needs a force of 200 N to push the car up the hill for 30 m. Path 2 needs a force of 600 N to pull the car up the hill 10 m. Show that the work is the same for both paths.

Step 1: Write the equation.

$$W_1 = F_1 \times d_1 \text{ and } W_2 = F_2 \times d_2$$

Step 2: Place values into the equation, and solve for the answer.

$$W_1 = 200 \text{ N} \times 30 \text{ m} \times 600 \text{ N and } W_2 = 600 \text{ N} \times 10 \text{ m} = 600 \text{ J}$$

The amount of work done is the same for both paths.

What Is Power?

The word *power* has a different meaning in science than how we often use the word. **Power** is how fast energy moves from one object to another.

Power measures how fast work is done. The *power output* of something is another way to say how much work can be done quickly. For example, a more powerful weightlifter can lift a barbell more quickly than a less powerful weightlifter.

How Is Power Calculated?

To calculate power (P), divide the work (W) by the time (t) it takes to do the work. This is shown in the following equation: $P = \frac{W}{t}$

Power is written in the units joules per second (J/s). This is called a **watt**. One *watt* (W) is the same as 1 J/s.

Let's do a problem. A stage manager at a play raises the curtain by doing 5,976 J of work on the curtain in 12 s. What is the power output of the stage manager?

Step 1: Write the equation.

$$P = \frac{W}{t}$$

Step 2: Place values into the equation, and solve for the answer.

$$P = \frac{5{,}976 \text{ J}}{12 \text{ s}} = 498 \text{ W}$$

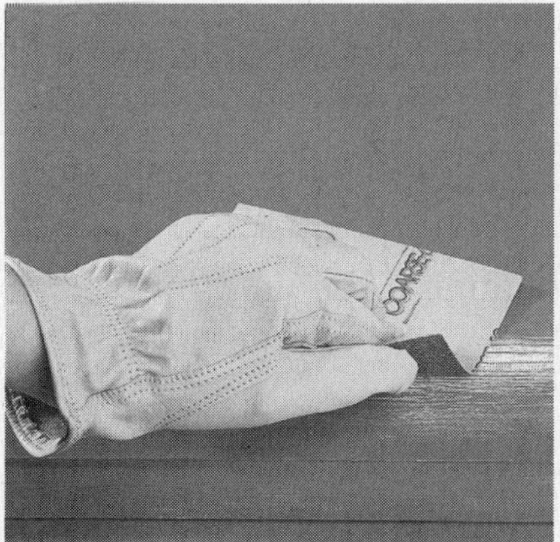

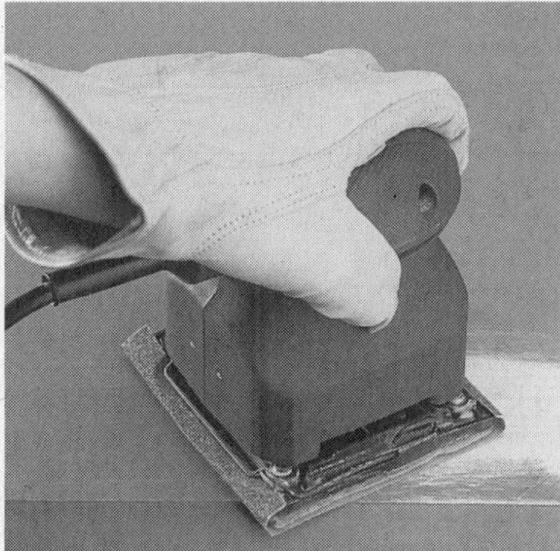

Wood can be sanded by hand or with an electric sander. The electric sander does the same amount of work faster.

Critical Thinking

8. Apply Concepts An escalator and a elevator can transport a person from one floor to the next. The escalator does it in 15 s and the elevator takes 10 s. Which does more work on the person? Which has the greater power output?

____

____

____

____

Math Focus

9. Calculate A light bulb is on for 12 s, and during that time it uses 1,200 J of electrical energy. What is the wattage (power output) of the light bulb?

TAKE A LOOK

10. Explain Why does the electric sander have a higher power output than sanding by hand?

____

____

____

Name ______________________ Class ______________ Date ______________

Section 1 Review

NSES PS 3a

SECTION VOCABULARY

joule the unit used to express energy; equivalent to the amount of work done by a force of 1 N acting through a distance of 1 m in the direction of the force (symbol, J)

power the rate at which work is done or energy is transformed

watt the unit used to express power; equivalent to a joule per second (symbol, W)

work the transfer of energy to an object by using a force that causes the object to move in the direction of the force

1. **Explain** Is work always done on an object when a force is applied to the object? Why or why not?

____

____

2. **Analyze** Work is done on a ball when a pitcher throws it. Is the pitcher still doing work on the ball as it flies through the air? Explain your answer.

____

____

3. **Calculate** A force of 10 N is used to push a shopping cart 10 m. How much work is done? Show how you got your answer.

4. **Compare** How is the term work different than the term power?

____

____

____

5. **Identify** How can you increase your power output by changing the amount of work that you do? How can you increase your power output by changing the time it takes you to do the work?

____

____

6. **Calculate** You did 120 J of work in 3 s. How much power did you use? Show how you got your answer.

Name ________________ Class ________________ Date ________________

SECTION 2 What Is a Machine?

BEFORE YOU READ

After you read this section, you should be able to answer these questions:

- What is a machine?
- How does a machine make work easier?
- What is mechanical advantage?
- What is mechanical efficiency?

What Is a Machine?

Imagine changing a flat tire without a jack to lift the car or a tire iron to remove the bolts. Would it be easy? No, you would need several people just to lift the car! Sometimes you need the help of machines to do work. A **machine** is something that makes work easier. It does this by lowering the size or direction of the force you apply. ☑

When you hear the word machine, what kind of objects do you think of? Not all machines are hard to use. You use many simple machines every day. Think about some of these machines. The following table lists some jobs you use a machine to do.

Work	Machine you could use
Removing the snow in your driveway	
Getting you to school in the morning	
Painting a room	
Picking up the leaves from your front yard	
Drying your hair	

Two Examples of Everyday Machines

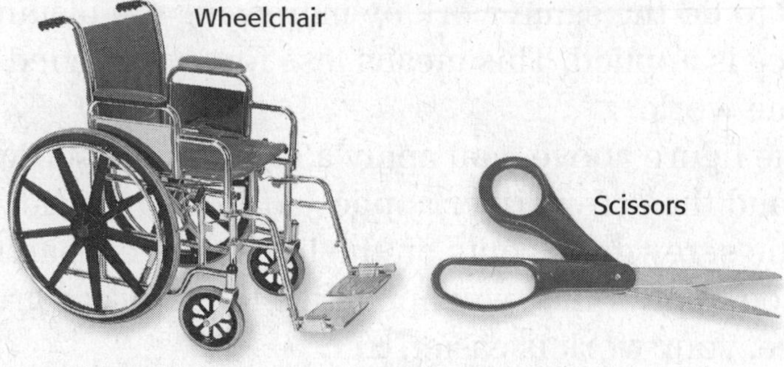

STUDY TIP

Brainstorm Think of ways that machines make your work easier and write them down in your science notebook.

READING CHECK

1. Describe How does a machine make work easier?

____

____

____

TAKE A LOOK

2. Identify Complete the table by filling in the last column.

How Do Machines Make Work Easier?

You can use a simple machine, such as a screwdriver, to remove the lid from a paint can. An example of this is shown in the figure below. The screwdriver is a type of *lever*. The tip of the screwdriver is put under the lid and you push down on the screwdriver. The tip of the screwdriver lifts the lid as you push down. In other words, you do work on the screwdriver, and the screwdriver does work on the lid.

WORK IN, WORK OUT

When you use a machine, you do work and the machine does work. The work you do on a machine is called the **work input**. The force you apply to the machine to do the work is the *input force*. The work done by the machine on another object is called the **work output**. The force the machine applies to do this work is the *output force*. ☑

READING CHECK

3. Identify What is work input? What is work output?

____

TAKE A LOOK

4. Identify What is the force you put on a screwdriver called?

____

What is the force the screwdriver puts on the lid called?

____

HOW MACHINES HELP

Machines do not decrease the amount of work you do. Remember that work equals the force applied times the distance ($W = F \times d$). Machines lower the force that is needed to do the same work by increasing the distance the force is applied. This means less force is needed to do the same work.

In the figure above, you apply a force to the screwdriver and the screwdriver applies a force to the lid. The force the screwdriver puts on the lid is greater than the force you apply. Since you apply this force over a greater distance, your work is easier. ☑

READING CHECK

5. Identify To make work easier, what force is lowered?

____

SAME WORK, DIFFERENT FORCE

Machines make work easier by lowering the size or direction (or both) of the input force. A machine doesn't change the amount of work done. A ramp can be used as a simple machine shown in the figure below. In this example a ramp makes work easier because the box is pushed with less force over a longer distance.

Input Force and Distance

The boy lifts the box. The input force is the same as the weight of the box.

The girl uses a ramp to lift the box. The input force is the less than the weight of the box. She applies this force for a longer distance.

Look at the boy lifting a box in the figure above. Suppose the box weighs 450 N and is lifted 1 m. How much work is done to move the box?

Step 1: Write the equation.

$$W = F \times d$$

Step 2: Place values into the equation, and solve.

$$W = 450 \text{ N} \times 1 \text{ m} = 450 \text{ N×m, or } 450 \text{ J}$$

Look at the girl using a ramp. Suppose the force to push the box is 150 N. It is pushed 3 m. How much work is done to move the box?

Step 1: Write the equation.

$$W = F \times d$$

Step 2: Place values into the equation, and solve.

$$W = 150 \text{ N} \times 3 \text{ m} = 450 \text{ N×m, or } 450 \text{ J}$$

Work done to move box is 450 J.

The same amount of work is done with or without the ramp. The boy uses more force and a shorter distance to lift the box. The girl uses less force and a longer distance to move the box. They each use a different force and a different distance to do the same work.

TAKE A LOOK

6. Describe Notice that the box is lifted the same distance by the boy and the girl. Which does more work on the box?

____

____

Math Focus

7. Calculate How much work is done when a 50 N force is applied to a 0.30 m screwdriver to lift a paint can lid?

FORCE AND DISTANCE CHANGE TOGETHER

When a machine changes the size of the output force, the distance must change. When the output force increases, the distance the object moves must decrease. This is shown in the figure of the nutcracker below. The handle is squeezed with a smaller force than the output force that breaks the nut. So, the output force is applied over a smaller distance.

Math Focus

8. Explain Why is the input arrow shorter than the output arrow in the photo of the nutcracker? Why are the arrows the same length in the figure of the pulley?

____

____

____

____

Machines Change the Size and/or Direction of a Force

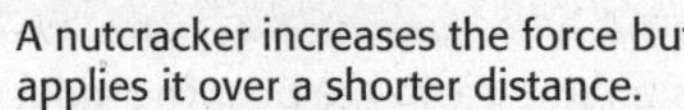

A nutcracker increases the force but applies it over a shorter distance.

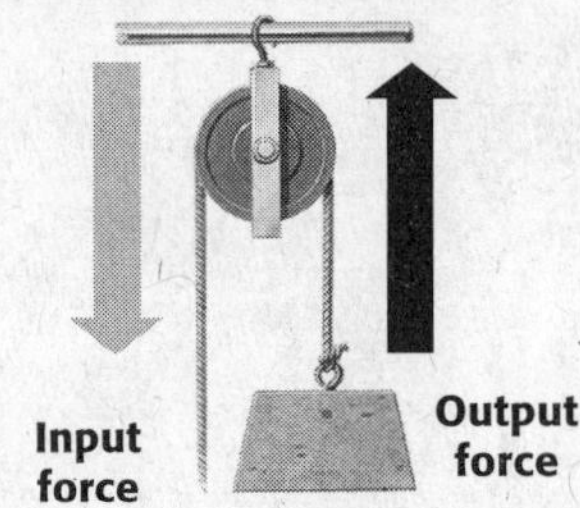

A simple pulley changes the direction of the input force, but the size of the output force is the same as the input force.

What Is Mechanical Advantage?

Some machines can increase the size of the force more than others. A machine's **mechanical advantage** tells you how much the force increases. The mechanical advantage compares the input force with the output force.

CALCULATING MECHANICAL ADVANTAGE

A machine's mechanical advantage can be calculated by using the following equation:

$$\textit{mechanical advantage (MA)} = \frac{\textit{output force}}{\textit{input force}}$$

Math Focus

9. Calculate What is the mechanical advantage of a nutcracker if the input force is 65 N and the output force is 130 N?

Look at this example. You push a box weighing 500 N up a ramp (output force) by applying 50 N of force. What is the mechanical advantage of the ramp?

Step 1: Write the equation.

$$\textit{mechanical advantage (MA)} = \frac{\textit{output force}}{\textit{input force}}$$

Step 2: Place values into the equation, and solve.

$$MA = \frac{500\text{ N}}{50\text{ N}} = 10$$

The mechanical advantage of the ramp is 10.

What Is a Machine's Mechanical Efficiency?

No machine changes all of the input work into output work. Some of the work done by the machine is lost to *friction*. Friction is always present when two objects touch. The work done by the machine plus the work lost to friction is equal to the work input. This is known as the *Law of Conservation of Energy*.

The **mechanical efficiency** of a machine compares a machine's work output with the work input. A machine is said to be efficient if it doesn't lose much work to friction.

CALCULATING MECHANICAL EFFICIENCY

A machine's mechanical efficiency is calculated using the following equation:

$$mechanical\ advantage\ (MA) = \frac{output\ force}{input\ force} \times 100$$

The 100 in the equation means that mechanical efficiency is written as a percentage. It tells you the percentage of work input that gets done as work output.

Let's try a problem. You do 100 J of work on a machine and the work output is 40 J. What is the mechanical efficiency of the machine?

Step 1: Write the equation.

$$mechanical\ effeciency\ (ME) = \frac{work\ output}{work\ input} \times 100$$

Step 2: Place values into the equation, and solve.

$$ME = \frac{40\text{ J}}{100\text{ J}} \times 100 = 40\%$$

Math Focus

10. Calculate What is the mechanical efficiency of a simple pulley if the input work is 100 N and the output work is 90 N?

Process Chart

You apply an input force to a machine.

↓

The machine changes the size and/or direction of the force.

↓

The machine applies an ________________ on the object.

↓

The mechanical efficiency and/or advantage of the machine can be determined.

Math Focus

11. Identify Fill in the missing words on the process chart.

____

Name ______________________ Class ______________ Date ______________

Section 2 Review

SECTION VOCABULARY

machine a device that helps do work by either overcoming a force or changing the direction of the applied force

mechanical advantage a number that tells how many times a machine multiplies force

mechanical efficiency a quantity, usually expressed as a percentage, that measures the ratio of work output to work input in a machine

work input the work done on a machine; the product of the input force and the distance through which the force is exerted

work output the work done by a machine; the product of the output force and the distance through which the force is exerted

1. **Explain** Why is it easier to move a heavy box up a ramp than it is to lift the box off the ground?

____

____

2. **Identify** What are the two ways that a machine can make work easier?

____

3. **Compare** What is the difference between work input and work output?

____

____

4. **Calculate** You apply an input force of 20 N to a hammer that applies an output force of 120 N to a nail. What is the mechanical advantage of the hammer? Show your work.

5. **Explain** Why is a machine's work output always less than the work input?

____

6. **Calculate** What is the mechanical efficiency of a machine with a work input of 75 J and a work output of 25 J? Show your work.

SECTION 3

Types of Machines

BEFORE YOU READ

After you read this section, you should be able to answer these questions:

- What are the six simple machines?
- What is a compound machine?

What Are the Six Types of Simple Machines?

All machines are made from one or more of the six simple machines. They are the lever, the pulley, the wheel and axle, the inclined plane, the wedge, and the screw. They each work differently to change the size or direction of the input force.

STUDY TIP As you read through the section, study the figures of the types of machines. Make a list of the six simple machines and a sentence describing how each works.

What Is a Lever?

A commonly used simple machine is the **lever**. A *lever* has a bar that rotates at a fixed point, called a *fulcrum*. The force that is applied to the lever is the *input force*. The object that is being lifted by the lever is called the *load*. A lever is used to apply a force to move a load. There are three classes of levers. They all have a different location for the fulcrum, the load, and the input force on the bar. ☑

READING CHECK

1. Describe How does a lever do work?

____

____

____

FIRST-CLASS LEVERS

In first-class levers, the fulcrum is between the input force and the load as shown in the figure below. The direction of the input force always changes in this type of lever. They can also be used to increase either the force or the distance of the work.

Examples of First-Class Levers

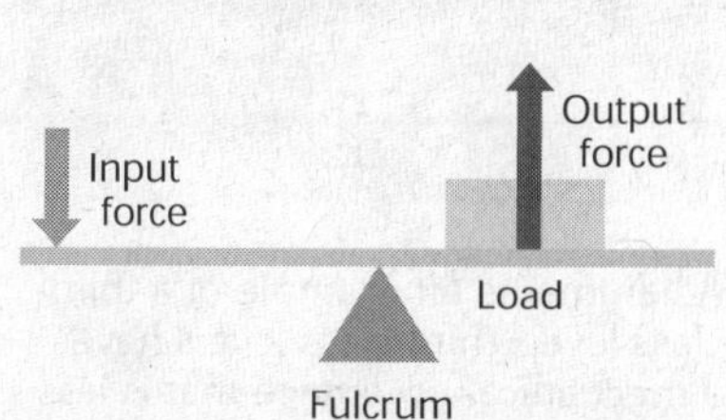

The fulcrum can be located closer to the load than to the input force. This lever has a mechanical advantage that is greater than 1. The output force is larger than the input force.

The fulcrum can be located exactly in the middle. This lever has a mechanical advantage that is equal to 1. The output force is the same as the input force.

Critical Thinking

2. Predict Suppose the fulcrum in the figure to the far left is located closer to the input force. How will this change the mechanical advantage of the lever? Explain.

____

____

____

____

Critical Thinking

3. Explain How does a second-class lever differ from a first-class lever?

____

____

____

____

____

SECOND-CLASS LEVERS

In second-class levers, the load is between the fulcrum and the input force as shown in the figure below. They do not change the direction of the input force. Second-class levers are often used to increase the force of the work. You apply less force to the lever than the force it puts on the load. This happens because the force is applied over a larger distance.

Examples of Second-Class Levers

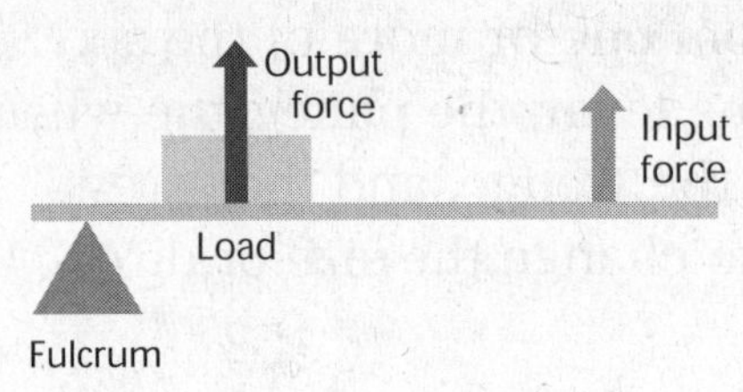

In a second-class lever, the output force, or load, is between the input force and the fulcrum.

A wheelbarrow is an example of a second-class lever. Second-class levers have a mechanical advantage that is greater than 1.

THIRD-CLASS LEVERS

In third-class levers, the input force is between the fulcrum and the load as shown in the figure below. The direction of the input force does not change and the input force does not increase. This means the output force is always less than the input force. Third-class levers do increase the distance that the output force works.

Critical Thinking

4. Explain Why can't a third-class lever have a mechanical advantage of 1 or more?

____

____

____

Examples of Third-Class Levers

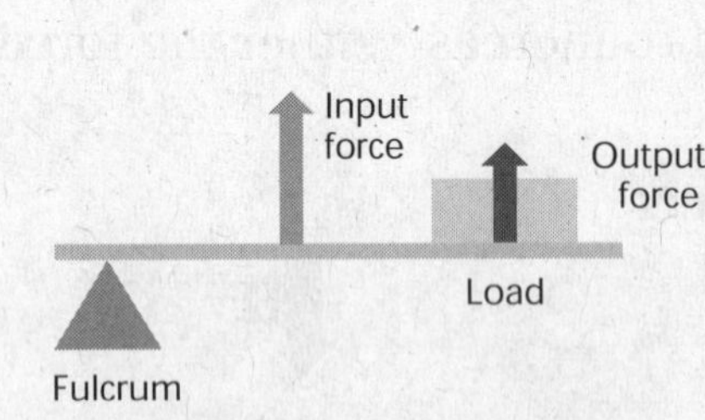

In a third-class lever, the input force is between the fulcrum and the load.

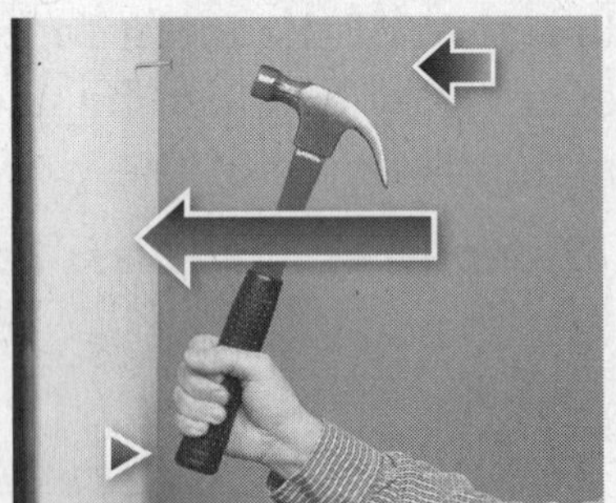

A hammer is an example of a third-class lever. Third-class levers have a mechanical advantage that is less than 1. The output force is less than the input force. Third-class levers increase the distance that the output force acts on.

What Is a Pulley?

When you open window blinds by pulling on a cord, you are using a pulley. A **pulley** is a simple machine with a grooved wheel that holds a rope or a cable. An input force is applied to one end of the cable. The object being lifted is called the load. The load is attached to the other end. The different types of pulleys are shown in the figure at the bottom of the page. ☑

READING CHECK

5. Describe What is a pulley?

FIXED PULLEYS

A *fixed pulley* is connected to something that does not move, such as a ceiling. To use a fixed pulley, you pull down on the rope to lift the load. The direction of the force changes. Since the size of the output force is the same as the input force, the mechanical advantage (*MA*) is 1. An elevator is an example of a fixed pulley. ☑

READING CHECK

6. Describe Why can't a fixed pulley have a mechanical advantage greater than 1?

MOVABLE PULLEYS

Moveable pulleys are connected directly to the object that is being moved, which is the load. The direction does not change, but the size of the force does. The mechanical advantage (*MA*) of a movable pulley is 2. This means that less force is needed to move a heavier load. Large construction cranes often use movable pulleys.

BLOCK AND TACKLES

If you use a fixed pulley and a movable pulley together, you form a pulley system. This is a *block and tackle*. The mechanical advantage (*MA*) of a block and tackle is equal to the number of sections of rope in the system.

Types of Pulleys

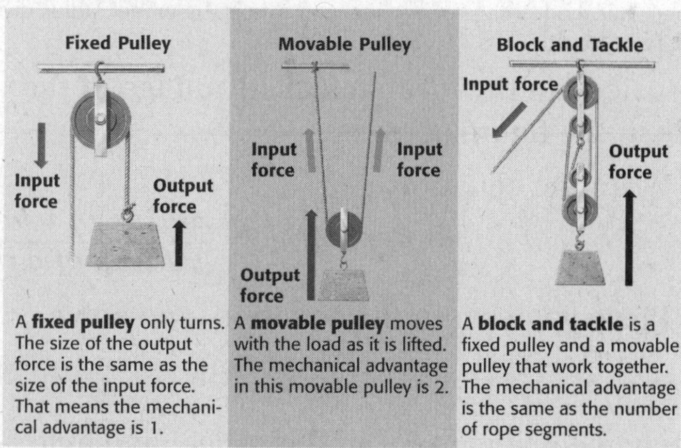

A **fixed pulley** only turns. The size of the output force is the same as the size of the input force. That means the mechanical advantage is 1.

A **movable pulley** moves with the load as it is lifted. The mechanical advantage in this movable pulley is 2.

A **block and tackle** is a fixed pulley and a movable pulley that work together. The mechanical advantage is the same as the number of rope segments.

TAKE A LOOK

7. Identify The section of rope labeled Input force for the block and tackle is not counted as a rope segment. There are four rope segments in this block and tackle. What is the mechanical advantage of the block and tackle?

Critical Thinking

8. Explain If the input force remains constant and the wheel is made smaller, what happens to the output force?

____

____

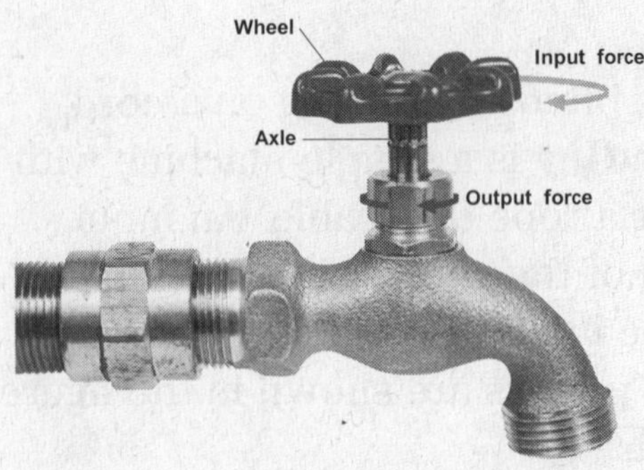

a When a small input force is applied to the wheel, it turns in a circular distance.

b When the wheel turns, so does the axle. The axle is smaller than the wheel. Since the axle turns a smaller distance, the output force is larger than the input force.

What Is a Wheel and Axle?

Did you know that a faucet is a machine? The faucet in the figure above is an example of a **wheel and axle**. It is a simple machine that is made up of two round objects that move together. The larger object is the *wheel* and the smaller object is the *axle*. Some examples of a wheel and axle are doorknobs, wrenches, and steering wheels.

MECHANICAL ADVANTAGE OF A WHEEL AND AXLE

The mechanical advantage (*MA*) of a wheel and axle can be calculated. To do this you need to know the *radius* of both the wheel and the axle. Remember, the radius is the distance from the center to the edge of the round object. The equation to find the mechanical advantage (*MA*) of a wheel and axle is:

$$\textit{mechanical advantage (MA)} = \frac{\textit{radius of wheel}}{\textit{radius of axle}}$$

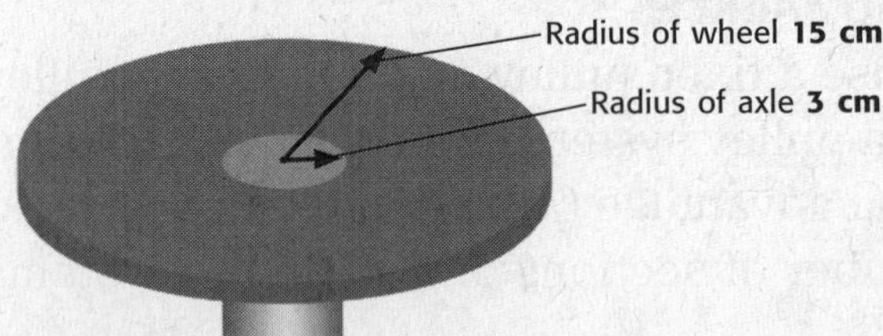

The mechanical advantage of a wheel and axle is the wheel radius divided by the axle radius.

Math Focus

9. Calculate A car has a wheel and axle. If the radius of the axle is 7.5 cm and the radius of the wheel is 75 cm, what is the mechanical advantage? Show your work.

Let's calculate the mechanical advantage of the wheel and axle in the figure above.

Step 1: Write the equation.

$$\textit{mechanical advantage (MA)} = \frac{\textit{radius of wheel}}{\textit{radius of axle}}$$

Step 2: Place values into the equation, and solve.

$$\text{MA} = \frac{15 \text{ cm}}{3 \text{ cm}} = 5$$

The mechanical advantage of this wheel and axle is 5.

What Is an Inclined Plane?

The Egyptians built the Great Pyramid thousands of years ago using the **inclined plane**. An *inclined plane* is a simple machine that is a flat, slanted surface. A ramp is an example of an inclined plane.

Using an inclined plane to move a heavy object into a truck is easier than lifting the object. The input force is smaller than the object's weight. The same work is done, but it happens over a longer distance. ☑

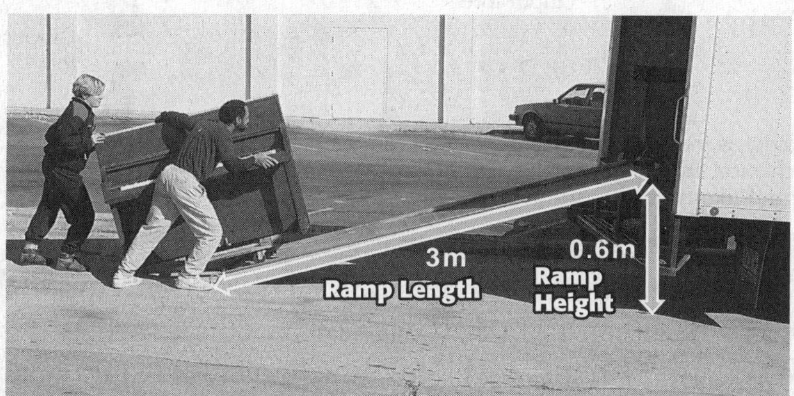

You do work to push a piano up a ramp. This is the same amount of work you would do to lift it straight up. An inclined plane lets you apply a smaller force over a greater distance.

READING CHECK

10. Explain How does an incline plane make lifting an object easier?

____
____
____
____

MECHANICAL ADVANTAGE OF INCLINED PLANES

The mechanical advantage (*MA*) of an inclined plane can also be calculated. The length of the inclined plane and the height the object that is lifted must be known. The equation to find the mechanical advantage (*MA*) of an inclined plane is:

$$\textit{mechanical advantage (MA)} = \frac{\textit{length of inclined plane}}{\textit{height load raised}}$$

We can calculate the mechanical advantage (*MA*) of the inclined plane shown in the figure above.

Step 1: Write the equation.

$$\textit{mechanical advantage (MA)} = \frac{\textit{length of inclined plane}}{\textit{height load raised}}$$

Step 2: Place values into the equation, and solve for the answer.

$$\text{MA} = \frac{3\text{ m}}{0.6\text{m}} = 5$$

The mechanical advantage (*MA*) is 5.

If the length of the inclined plane is much greater than the height, the mechanical advantage is large. That means an inclined plane with a gradual slope needs less force to move objects than a steep-sloped one.

Math Focus

11. Determine An inclined plane is 10 m and lifts a piano 2.5 m. What is the mechanical advantage of the inclined plane? Show your work.

What Is a Wedge?

A knife is often used to cut because it is a **wedge**. A *wedge* is made of two inclined planes that move. Like an inclined plane, a wedge needs a small input force over a large distance. The output force of the wedge is much greater than the input force. Some useful wedges are doorstops, plows, ax heads, and chisels.

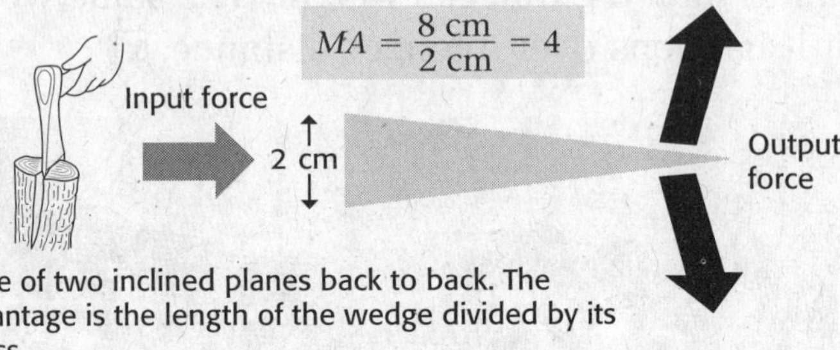

A wedge is made of two inclined planes back to back. The mechanical advantage is the length of the wedge divided by its greatest thickness.

TAKE A LOOK

12. Predict What would happen to the mechanical advantage of the wedge if it were longer in length?

____

MECHANICAL ADVANTAGE OF WEDGES

The mechanical advantage of a wedge can be found by dividing the length of the wedge by its greatest thickness. The equation to find a wedge's mechanical advantage is:

$$\textit{mechanical advantage (MA)} = \frac{\textit{length of wedge}}{\textit{largest thickness of wedge}}$$

A wedge has a greater mechanical advantage if it is long and thin. When you sharpen a knife you are making the wedge thinner. This needs a smaller input force.

What Is a Screw?

A **screw** is an inclined plane that is wrapped around a cylinder. To turn a screw, a small force over a long distance is needed. The screw applies a large output force over a short distance. Screws are often used as fasteners.

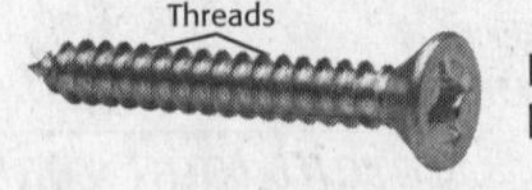

If you could unwind a screw, you would have a very long inclined plane.

Say It

Demonstrate Take a long pencil and a piece of paper cut so it looks like an inclined plane. Roll the paper around the pencil so it looks like threads on a screw. Show the class how the paper looks like threads on a screw. Then unwind the paper showing that it looks like an inclined plane.

MECHANICAL ADVANTAGE OF SCREWS

To find the mechanical advantage of a screw you need to first unwind the inclined plane. Then, if you compare the length of the inclined plane with its height you can calculate the mechanical advantage. This is the same as calculating the mechanical advantage of an inclined plane. The longer the spiral on a screw and the closer the threads, the greater the screw's mechanical advantage.

What Is a Compound Machine?

There are machines all around you. Many machines do not look like the six simple machines that you have read about. That is because most of the machines in the world are **compound machines**. These are machines that are made of two or more simple machines. A block and tackle is one example of a compound machine that you have already seen. It is made of two or more pulleys. ☑

A common example of a compound machine is a can opener. A can opener may look simple, but it is made of three simple machines. They are the second-class lever, the wheel and axle, and the wedge. When you squeeze the handle, you are using a second-class lever. The blade is a wedge that cuts the can. When you turn the knob to open the can, you are using a wheel and axle.

READING CHECK

13. Describe What is a compound machine?

____

____

____

A can opener is a compound machine. The handle is a second-class lever, the knob is a wheel and axle, and a wedge is used to open the can.

TAKE A LOOK

14. Describe Describe the process of using a can opener. Tell the order in which each simple machine is used and what it does to open the can.

____

____

____

____

____

____

MECHANICAL EFFICIENCY OF COMPOUND MACHINES

The *mechanical efficiency* of most compound machines is low. Remember that mechanical efficiency tells you what percentage of work input gets done as work output. This is different than the mechanical advantage. The efficiency of compound machines is low because they usually have many moving parts. This means that there are more parts that contact each other and more friction. Recall that friction lowers output work. ☑

Cars and airplanes are compound machines that are made of many simple machines. It is important to lower the amount of friction in these compound machines. Friction can often damage machines. Grease is usually added to cars because it lowers the friction between the moving parts.

READING CHECK

15. Identify Why do most compound machines have low mechanical efficiency?

____

____

____

Name ______________________ Class ______________ Date ______________

Section 3 Review

SECTION VOCABULARY

compound machine a machine made of more than one simple machine **inclined plane** a simple machine that is a straight, slanted surface, which facilitates the raising of loads; a ramp **lever** a simple machine that consists of a bar that pivots at a fixed point called a fulcrum **pulley** a simple machine that consists of a wheel over which a rope, chain, or wire passes	**screw** a simple machine that consists of an inclined plane wrapped around a cylinder **wedge** a simple machine that is made up of two inclined planes and that moves; often used for cutting **wheel and axle** a simple machine consisting of two circular objects of different sizes; the wheel is the larger of the two circular objects

1. Compare Use a Venn Diagram to compare a first-class lever and a second-class lever.

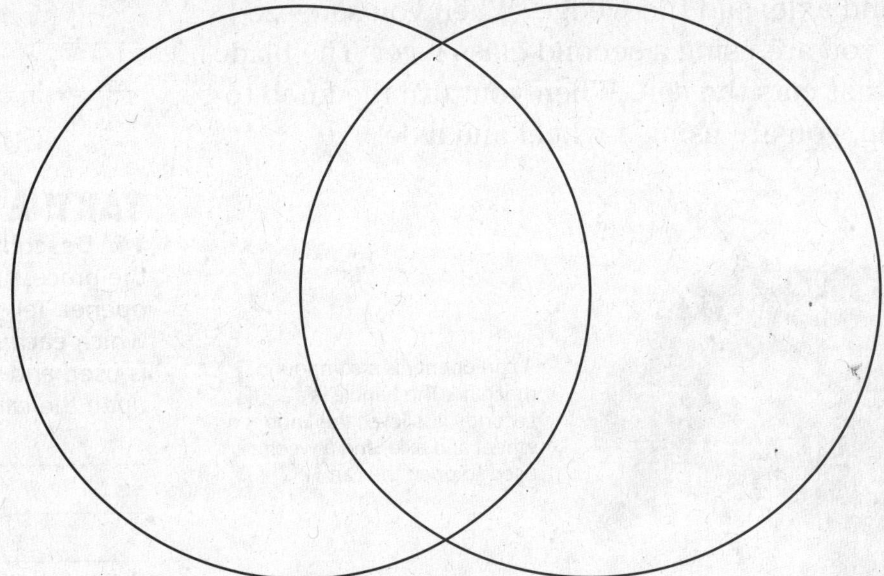

2. Calculate A screwdriver is used to put a screw into a piece of wood. The radius of the handle is 1.8 cm and the radius of shaft is 0.6 cm. What is the mechanical advantage of using the screwdriver? Show your work.

3. Compare What is the difference between a wedge and a screw?

____

____

4. Analyze When there is a lot of friction in a machine, what is lowered and causes mechanical efficiency to be lowered?

____

Name ______________________ Class ______________ Date ______________

CHAPTER 5 Energy and Energy Resources

SECTION 1

What Is Energy?

BEFORE YOU READ

After you read this section, you should be able to answer these questions:

- How are energy and work related?
- How is kinetic energy different from potential energy?
- What are some of the other forms of energy?

National Science Education Standards

PS 3a, 3d, 3e, 3f

What Is Energy?

A tennis player needs energy to hit a ball with her racket. The ball has energy as it flies through the air. Energy is all around you, but what is energy?

In science, **energy** is the ability to do work. *Work* is done when a force makes an object move in the direction of the applied force. How do energy and work help you play tennis? The tennis player does work on her racket by applying a force to it. The racket does work on the ball to make it fly into the air.

When the racket does work on the ball, energy moves from the racket to the ball. Energy is the reason the racket can do work. So, work is the transfer of energy. Both work and energy are written in the units joules (J). ☑

STUDY TIP

Make a Venn Diagram to compare and contrast kinetic energy and potential energy. Make a list of other forms of energy and tell if they are kinetic energy or potential energy.

READING CHECK

1. Identify When work is done by one object on another, what is transferred?

____

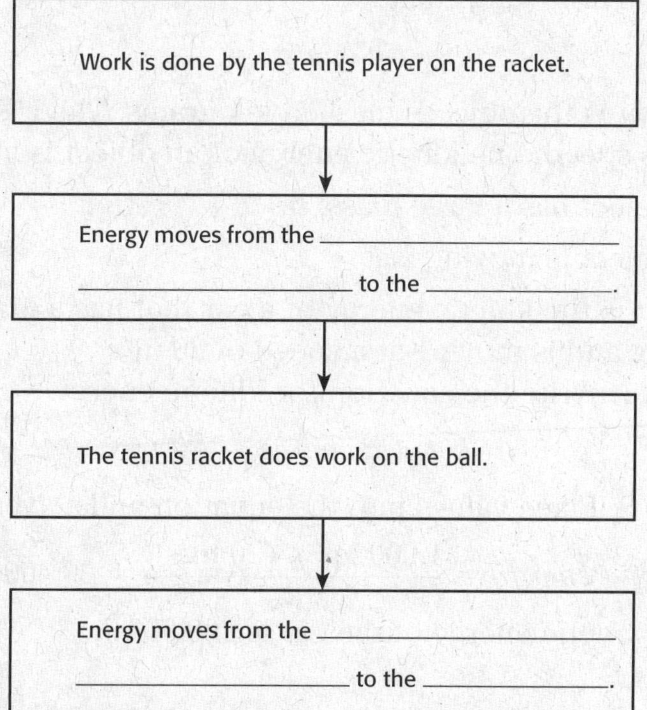

TAKE A LOOK

2. Identify Fill in the process chart to show how energy moves when work is done on an object.

What Is Kinetic Energy?

When the tennis player hits the ball with the racket, energy moves from the racket to the ball. The tennis ball has kinetic energy. **Kinetic energy** is the energy of motion. All moving objects have kinetic energy. Like all other forms of energy, kinetic energy can be used to do work. The kinetic energy of a hammer does work on a nail. This is seen in the figure below. ☑

READING CHECK

3. Identify For an object to have kinetic energy, what must it be doing?

____

When you swing a hammer, it is moving. It has kinetic energy. This energy does work on the nail driving it into the wood.

CALCULATING KINETIC ENERGY

You can calculate the kinetic energy of an object by using the following equation.

$$kinetic\ energy = \frac{mv^2}{2}$$

The m is the object's mass in kilograms. The v is the object's speed. The kinetic energy of an object is large if:

- the object has a large mass, or
- the object is moving fast

What is the kinetic energy of a car that has a mass of 1,000 kg and is moving at a speed of 20 m/s?

Step 1: Write the equation for kinetic energy.

$$kinetic\ energy = \frac{mv^2}{2}$$

Step 2: Place values into the equation and solve.

$$kinetic\ energy = \frac{1{,}000\ \text{kg} \times (20\ \text{m/s})^2}{2} = 200{,}000\ \text{J}$$

The kinetic energy of the car is 200,000 J.

Math Focus

4. Calculate What is the kinetic energy of a 0.50 kg hammer that hits the floor at a speed of 10 m/s?

What Is Potential Energy?

An object does not have to be moving to have energy. **Potential energy** is the energy an object has because of its position. This kind of energy is harder to see because we do not see the energy at work. In the figure below, when the bow is pulled back, it has potential energy. Work has been done on it, and that work has been turned into potential energy. ☑

READING CHECK

5. Identify What causes an object to have potential energy?

____

The bow and the string have energy that is stored as potential energy. When the man lets go of the string, the potential energy does work on the arrow.

Critical Thinking

6. Infer What would the man need to do to give the arrow more potential energy?

____

____

GRAVITATIONAL POTENTIAL ENERGY

When you lift an object, you do work on it. You move it in an opposite direction from the force of gravity. As you lift the object, you transfer energy to the object and give it gravitational potential energy. The amount of *gravitational potential energy* of an object depends on the object's weight and its distance from the ground.

CALCULATING GRAVITATIONAL POTENTIAL ENERGY

The gravitational potential energy of an object can be determined by using the following equation:

gravitational potential energy = *weight* × *height*

The weight is in newtons (N) and the height is in meters (m). Gravitational potential energy is written in newton × meters (N × m). This is the same as a joule (J).

Let's do a calculation. What is the gravitational potential energy of a book with a weight of 13 N at a height of 1.5 m off the ground?

Step 1: *gravitational potential energy* = *weight* × *height*

Step 2: *gravitational potential energy* = 13 N × 1.5 m = 19.5 J

The book now has 19.5 J of potential energy.

Math Focus

7. Calculate What is the potential energy of a 300 N rock climber standing 100 m from the base of a rock wall?

What Is Mechanical Energy?

Look at the figure below. All the energy in the juggler's pins is in the form of mechanical energy. **Mechanical energy** is the total energy of motion and position of an object. In other words, it is the kinetic energy plus the potential energy of an object. ☑

READING CHECK

8. Identify Adding what two energies gives the mechanical energy of an object?

____

____

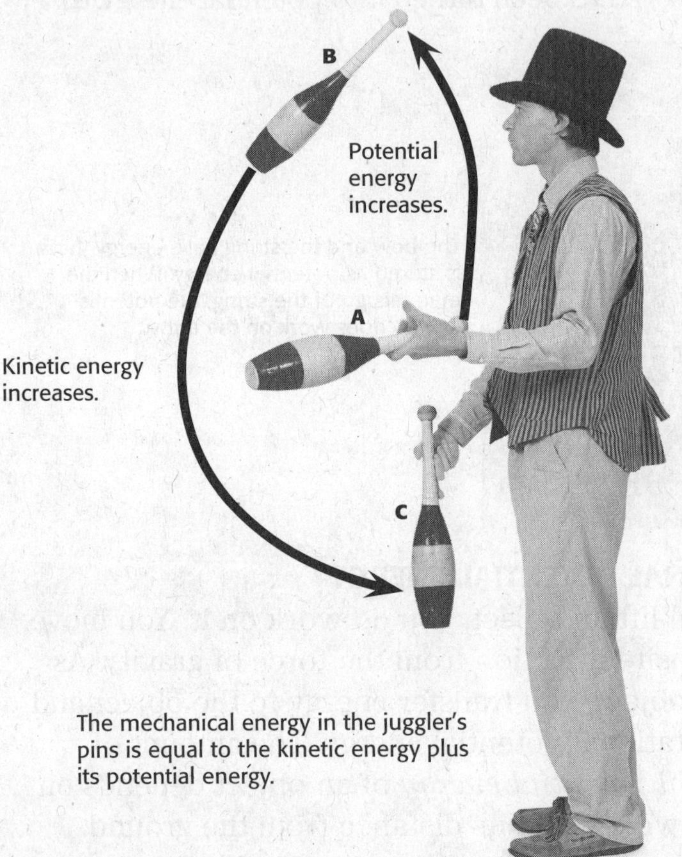

The mechanical energy in the juggler's pins is equal to the kinetic energy plus its potential energy.

TAKE A LOOK

9. Identify In the figure, circle the pin with the most potential energy.

Discuss Suppose a batter hits a pop-up in baseball that goes straight up. With a partner, discuss the changes in kinetic and potential energy as the ball leaves the bat and rises to its highest height.

MECHANICAL ENERGY IN A JUGGLER'S PIN

The mechanical energy of an object doesn't change unless energy is transferred to or from another object.

Look again at the figure of the juggler. The juggler moves the pin by doing work on it. He gives the pin kinetic energy. When he lets go of the pin, the pin's kinetic energy changes into potential energy. As the pin goes up, it slows down. When all of the pin's kinetic energy is turned into potential energy, it stops going up.

When the pin starts to fall, its energy is mostly potential energy. As it falls, the potential energy is changed back into kinetic energy. At different times, the pin may have more kinetic energy or more potential energy. The total mechanical energy at any point is always the same.

What Are the Other Forms of Energy?

Energy can be in a form other than mechanical energy. The other energy forms are thermal, chemical, electrical, sound, light, and nuclear energy. All of these energy forms are connected in some way to kinetic energy and potential energy.

THERMAL ENERGY

Matter is made of particles that are moving. These particles have kinetic energy. *Thermal energy* is all the kinetic energy from the movement of the particles in an object. ☑

The figure below shows the thermal energy of particles at different temperatures. Particles move faster at higher temperatures than at lower temperatures. The faster the particles move, the greater their kinetic energy and thermal energy are.

READING CHECK

10. Describe What is thermal energy?

____

____

____

The Thermal Energy in Water

The particles in an ice cube vibrate in fixed positions and do not have a lot of kinetic energy.

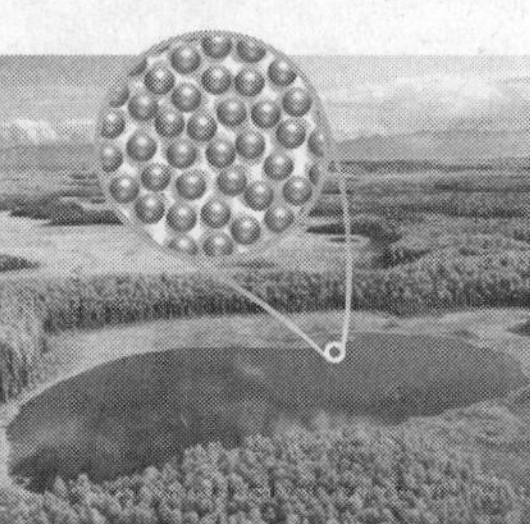

The particles in water in a lake can move more freely and have more kinetic energy than water particles in ice do.

The particles in water in steam move rapidly, so they have more energy than particles in liquid water do.

CHEMICAL ENERGY

Chemical compounds such as sugar, salt, and water store energy. These compounds are made of many atoms that are held together by chemical bonds. Work is done to join the atoms together to form these bonds. *Chemical energy* is the energy stored in the chemical bonds that hold the compounds together. Chemical energy is a type of potential energy because it depends on the position of the atoms in the compound. ☑

11. Identify Chemical energy is a form of what type of energy?

____

STANDARDS CHECK

PS 3d Electrical circuits provide a means of transferring electrical energy when heat, light, sound, and chemical changes are produced.

12. Describe How does an amplifier make sound?

____

____

____

____

ELECTRICAL ENERGY

You use electrical energy every day. Electrical outlets in your home allow you to use this energy. *Electrical energy* is the energy of moving particles called electrons. Electrons are the negatively charged particles of atoms.

What happens when you plug an electrical device, such as an amplifier shown in the figure below, into an outlet? You use electrical energy. The electrons in the wires move to the amplifier. The moving electrons do work on the speaker in the amplifer. This makes the sound that you hear from the amplifier.

Electrical energy has both kinetic energy and potential energy. When electrical energy runs through a wire, it uses its kinetic energy. Electrical energy that is waiting to be used is potential energy. This potential energy is in the wire before you plug in an electrical appliance.

TAKE A LOOK

13. Identify What is the energy source for the amplifier? What kind of energy is transmitted by the guitar?

____

____

____

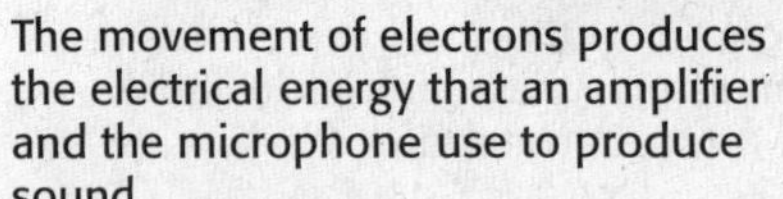

The movement of electrons produces the electrical energy that an amplifier and the microphone use to produce sound.

As the guitar strings vibrate, they cause particles in the air to vibrate. These vibrations transmit sound energy.

SOUND ENERGY

Sound energy is the energy from a vibrating object. *Vibrations* are small movements of particles of an object. In the figure above, the guitar player pulls on the guitar string. This gives the string potential energy. When she lets go of the string, the potential energy turns into kinetic energy. This makes the string vibrate.

When the guitar string vibrates, some of its kinetic energy moves to nearby air particles. These vibrating air particles cause sound energy to travel. When the sound energy reaches your ear, you hear the sound of the guitar.

LIGHT ENERGY

Light helps you see, but not all light can be seen. We use light in microwaves, but we do not see it. *Light energy* is made from vibrations of electrically charged particles. Light energy is like sound energy. They both happen because particles vibrate. However, light energy doesn't need particles to travel. This makes it different than sound energy. Light energy can move through a *vacuum*, which is an area where there is no matter.

Microwave Oven

The energy used to cook food in a microwave is a form of

_______________ _______________.

TAKE A LOOK

14. Identify Complete the sentence found in the figure.

NUCLEAR ENERGY

Another kind of energy is stored in the nucleus of an atom. This energy is *nuclear energy*. This energy is stored as potential energy.

There are two ways nuclear energy can be given off by a nucleus. When two or more small nuclei join together, they give off energy in a reaction called *fusion*. The sun's light and heat come from fusion reactions.

The second way nuclear energy is given off is when a nucleus splits apart. This process is known as *fission*. Large nuclei, like uranium, can be broken apart with fission. Fission is used to create electrical energy at nuclear power plants. ☑

READING CHECK

15. Compare How do nuclear fusion and fission differ?

____

____

____

Our Sun

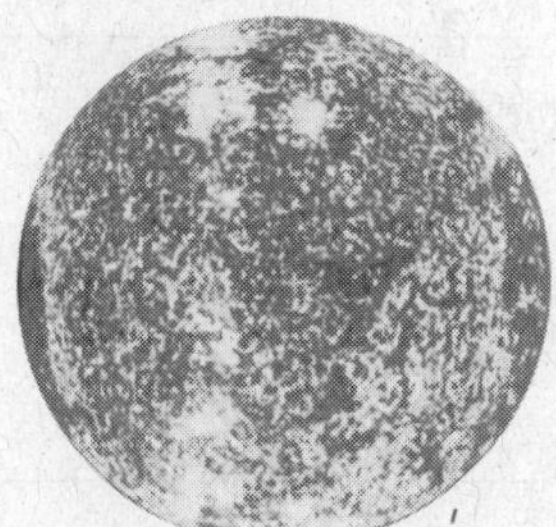

Without _______________

_______________ that gives the sun its energy, life on Earth would not be possible.

TAKE A LOOK

16. Identify Complete the sentence found in the figure.

Name ____________________ Class ____________ Date ____________

Section 1 Review

NSES PS 3a, 3d, 3e, 3f

SECTION VOCABULARY

energy the capacity to do work

kinetic energy the energy of an object that is due to the object's motion

mechanical energy the amount of work an object can do because of the object's kinetic and potential energies

potential energy the energy that an object has because of the position, shape, or condition of the object

1. **Identify** An object's kinetic energy depends on two things. What are they?

____

2. **Calculate** A book weighs 16 N and is placed on a shelf that is 2.5 m from the ground. What is the gravitational potential energy of the book? Show your work.

3. **Calculate** What is the kinetic energy of a 2,000 kg bus that is moving at 25 m/s? Show your work.

4. **Explain** A girl is jumping on a trampoline. When she is at the top of her jump, her mechanical energy is in what form? Explain why.

____

____

____

5. **Identify** How are sound energy and light energy similar?

____

____

6. **Conclude** Which type of nuclear reaction is more important for life on Earth? Explain why.

____

____

Name _______________ Class _______________ Date _______________

CHAPTER 5 Energy and Energy Resources

SECTION 2

Energy Conversions

BEFORE YOU READ

After you read this section, you should be able to answer these questions:

- What are energy conversions?
- How do machines use energy conversions?

National Science Education Standards

PS 3a, 3f

What Is an Energy Conversion?

Think of a book sitting on a shelf. The book has gravitational potential energy when it is on the shelf. What happens if the book falls off the shelf? Its potential energy changes into kinetic energy. This is an example of an energy conversion.

An **energy conversion** is a change from one form of energy to another. Any form of energy can change into any other form of energy. One form of energy can sometimes change into more than one other energy form.

STUDY TIP

Make a list in your science notebook of the energy conversions discussed in this section.

KINETIC AND POTENTIAL ENERGY CONVERSION

A common energy conversion happens between potential energy and kinetic energy. When the skateboarder in the figure below moves down the half-pipe, he has a lot of kinetic energy. As he travels up the half-pipe, his kinetic energy changes into potential energy.

Potential Energy and Kinetic Energy

At the bottom of the half-pipe, the skateboarder has a lot of speed. His _______________ _______________ is at a maximum.

TAKE A LOOK

1. Identify Complete the two blanks in the figure.

ELASTIC POTENTIAL ENERGY

Another example of potential energy changing into kinetic energy is shown in a rubber band. When you stretch a rubber band, you give it potential energy. This energy is called *elastic potential energy*. When you let go of the rubber band, it flies across the room. The stored potential energy in the stretched rubber band is turned into kinetic energy. ☑

READING CHECK

2. Describe How does a rubber band gain elastic potential energy?

____

How Do Chemical Energy Conversions Happen?

You may have heard that breakfast is the most important meal of the day. Why is eating breakfast so important? *Chemical energy* comes from the food you eat. Your body changes the chemical energy from food into several different energy forms that it can use. Some of these are:

- mechanical energy, to move your muscles
- thermal energy, to keep your body temperature constant
- electrical energy, to help your brain think

So eating breakfast helps your body do all of your daily activities.

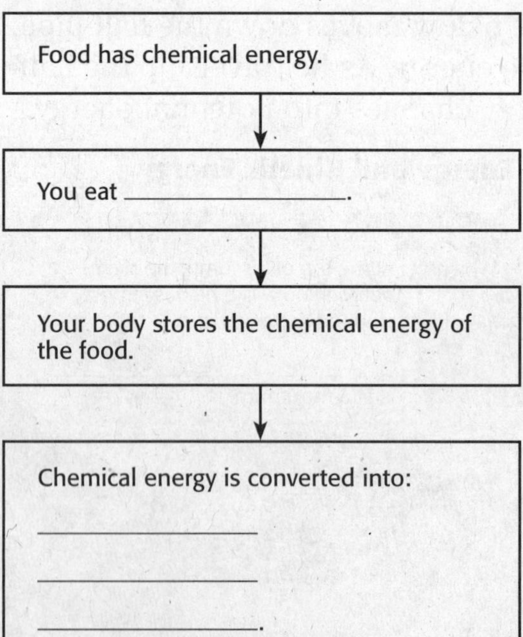

TAKE A LOOK

3. Identify Fill in the process chart to show how energy is converted in your body.

From Light Energy to Chemical Energy

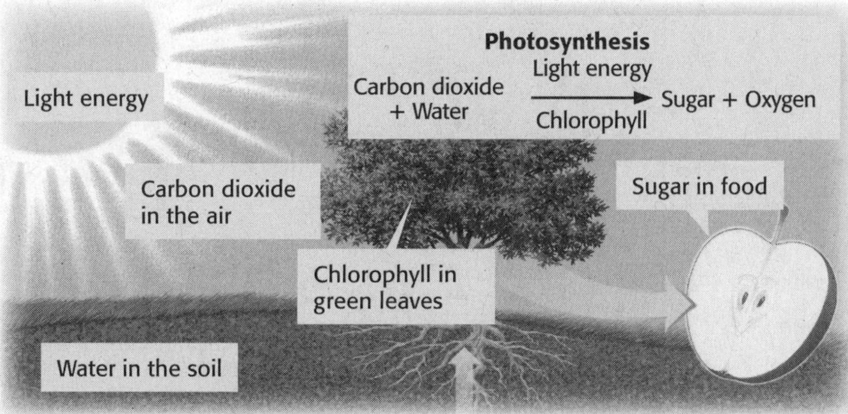

STANDARDS CHECK

PS 3f The sun is a major source of energy for changes on the earth's surface. The sun loses energy by emitting light. A tiny fraction of that light reaches the earth, transferring energy from the sun to the earth. The sun's energy arrives as light with a range of wave lengths, consisting of visible light, infrared, and ultraviolet radiation.

4. Identify Light energy from the sun is converted into what kind of energy by green leaves?

____

ENERGY CONVERSIONS IN PLANTS

Did you know the chemical energy in the food you eat comes from the sun's energy? When you eat fruits, vegetables, or grains, you are taking in chemical energy. Energy from the sun was used to make the chemical energy in the food. Many animals also eat plants. If you eat meat from these animals, you are also taking in energy that comes from the sun.

The figure above shows how light energy is used to make new material that has chemical energy. When photosynthesis happens in plants, light energy from the sun is changed into chemical energy. *Photosynthesis* is a chemical reaction in plants that changes light energy into chemical energy. When we eat fruits, vegetables, or grains, we are eating the chemical energy that is stored in the plants.

The chemical energy from a tree can also be changed into thermal energy. This change happens when you burn a tree's wood. If you go back far enough, you would see that the energy from a wood fire comes from the sun.

Discuss With a partner, describe how light energy from the sun becomes chemical energy in meat from animals.

Energy Conversions in a Hair Dryer

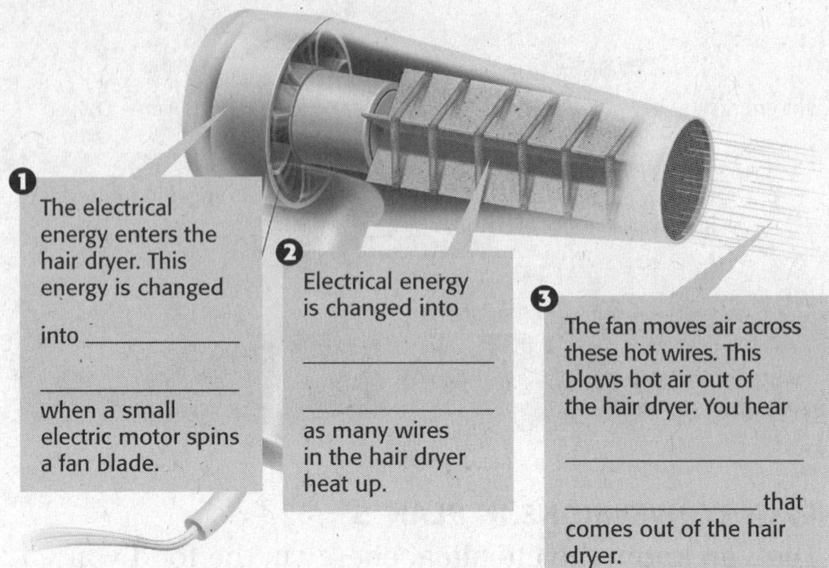

TAKE A LOOK
5. Identify Fill in the blanks in the figure.

Why Are Energy Conversions Important?

Energy conversions happen everywhere. Heating our homes and getting energy from food are just a few examples of how we use energy conversions. Machines also convert and use energy. This can be seen from the figure of the hair dryer above. Electrical energy by itself won't dry your hair. The hair dryer changes electrical energy into thermal energy. This heat helps dry your hair.

ELECTRICAL ENERGY CONVERSIONS

You use electrical energy all of the time. Some examples of when you use electrical energy are listening to the radio, making toast, and taking a picture. Electrical energy changes into other kinds of energy easily.

TAKE A LOOK
6. Identify Look at the table. In each example, think about what kind of energy conversion happens. Then, fill in the last column of the table with the kind (or kinds) of energy that is formed.

Some Conversions of Electrical Energy		
Alarm Clock	Electrical energy →	
Hair dryer	Electrical energy →	
Light bulb	Electrical energy →	
Blender	Electrical energy →	

How Do Machines Use Energy?

You have seen how energy can change into different forms. Another way to look at energy is to see how machines use energy. Remember that a machine can make work easier. It does this by changing the size or direction (or both) of the force that does the work. The machine converts the energy you put into the machine into work.

There is another way that machines can help you use energy. They can convert energy into a form that you need, such as a stove using electrical energy to cook food. ☑

An example of how energy is used by a machine is shown in the figure below. The biker puts a force on the pedals. This transfers kinetic energy from the biker into kinetic energy to move the pedals. This energy moves to other parts of the bike. The bike lets the biker use less force over a greater distance. This makes his work easier.

READING CHECK

7. Describe What are two ways a machine converts energy?

____

____

____

____

Energy Conversions in a Bicycle

For your bike to start and keep moving, energy must be transferred and converted.

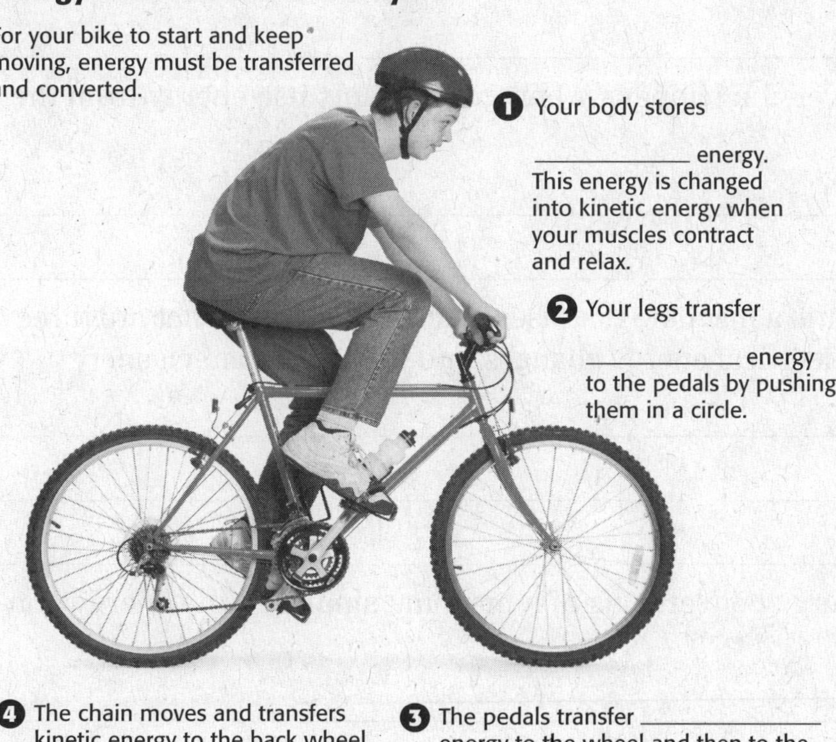

TAKE A LOOK

8. Identify Complete the blanks in the figure.

____

____

____

____

Name ______________________ Class ______________ Date ______________

Section 2 Review

NSES PS 3a, 3f

SECTION VOCABULARY

energy conversion a change from one form of energy to another	

1. **Identify** Suppose a skier is standing still at the top of a hill. What kind of energy does the skier have?

____

2. **Explain** What happens to the energy of the skier in Question 1 if he goes down the hill?

____

3. **Describe** How does your body get the energy that it needs to do all of its daily activities? Describe the energy conversions that take place.

____

____

____

____

4. **Explain** What energy conversion happens when green plants use energy from the sun to make sugar?

____

____

5. **Analyze** A vacuum cleaner is a machine that uses electrical energy. What are three forms of energy that the electrical energy changes into in the vacuum cleaner?

____

____

____

6. **Compare** How are the energy conversions of a machine similar to the energy conversions in your body?

____

____

____

Name ________________ Class ________________ Date ________________

SECTION 3

Conservation of Energy

BEFORE YOU READ

After you read this section, you should be able to answer these questions:

- How does friction affect energy conversions?
- What is the law of conservation of energy?
- What happens to energy in a closed system?

National Science Education Standards
PS 3a

Where Does Energy Go?

Most roller coasters use a chain to pull the cars up to the top of the first hill. As the cars go up and down the rest of the hills, their potential energy and kinetic energy keep changing. Since no additional energy is put into the cars, they never go as high as the first hill. Is energy lost? No, it just changes into other forms of energy.

When a roller coaster's energy is changing, some of its energy is lowered by friction. **Friction** is a force that is present when two objects touch each other. There is friction between the cars' wheels and the track. There is also friction between the cars and the air around them. Friction slows the motion of a roller coaster. ☑

STUDY TIP
In your science notebook, list the energy conversions in this section and note why energy is not conserved in the conversion.

READING CHECK

1. Identify What causes the energy of a roller coaster to be lowered when the coaster's energy changes form?

____

Energy Conversions in a Roller Coaster

Not all of the cars' potential energy (PE) changes into kinetic energy (KE) as the cars go down the first hill. When the cars move up hill 2, not all of its KE is converted into PE at the top of hill 2. Some of it is changed into thermal energy because of friction.

TAKE A LOOK

2. Complete Fill in the blanks in the figure with the following terms (they can be used more than once): greater, less, largest, smallest.

What Happens to Energy in a Closed System?

A *closed system* is a group of objects that move energy only to each other. On a roller coaster, a closed system would be the track, the cars, and the air around the cars.

Some of the mechanical energy (kinetic energy plus potential energy) of the cars changes into thermal energy. This happens because of friction. Mechanical energy is also converted into sound energy. You hear this energy when you are near the roller coaster. The rest of the potential energy changes into kinetic energy. This is seen by the roller coaster racing down the track.

If you add up all of this energy, it equals the cars' potential energy at the top of the first hill. The energy at the top of the hill is the same as the energy that is converted to other forms. So, energy is conserved and not lost in a closed system.

STANDARDS CHECK

PS 3a Energy is a property of many substances and is associated with heat, light, electricity, mechanical motion, sound, nuclei, and the nature of a chemical. Energy is transferred in many ways.

3. Identify Into what form of energy does friction convert mechanical energy?

____

LAW OF CONSERVATION OF ENERGY

Energy is always conserved. This is always true, so it is called a law. In the **law of conservation of energy**, energy cannot be created or destroyed. In a closed system, energy can change from one form to another, but the total energy is always the same. It doesn't matter how many or what kinds of energy conversions take place. The energy conversions that happen in a light bulb are shown in the figure below. ☑

READING CHECK

4. Describe What is the law of conservation of energy?

____

____

Energy Conservation in a Light Bulb

Some energy is converted into thermal energy, which makes the bulb feel warm.

Some electrical energy is converted into light energy.

As electrical energy is carried through the wires, some of it changes into thermal energy.

Critical Thinking

5. Explain Why is the light bulb *not* a closed system?

____

____

____

____

Why Can't a Machine Run Forever Without Adding Energy?

Any time an energy conversion happens, some of the energy always changes into thermal energy. This thermal energy is in the form of friction. The energy from friction is not useful energy. It is not used to do work.

Think about a car. You put gas into a car. Not all of the chemical energy from the gas makes the car move. Some energy is wasted as thermal energy. This energy leaves through the radiator and the exhaust pipe.

It is impossible to create a machine that runs forever without adding more energy to it. This kind of machine is called a *perpetual motion machine*. A machine has to have a constant supply of energy because energy conversions always produce wasteful thermal energy. ☑

6. Describe Why can't there be a perpetual motion machine?

____

____

EFFICIENT ENERGY CONVERSIONS

If a car gets good gas mileage, it is energy efficient. The *energy efficiency* is found by comparing the amount of starting energy with the amount of useful energy produced. A car with high energy efficiency goes farther than other cars with the same amount of gasoline.

More efficient energy conversions waste less energy. For example, smooth, *aerodynamic* cars have less friction between the car and the air around it. They use less energy, so they are more efficient. If less energy is wasted, then less energy is needed to run the car.

Car A

The shape of newer cars lowers the friction between the car and the air passing over it.

Car B

READING CHECK

7. Describe Which car is more aerodynamic in shape?

____

Name ____________________ Class ____________ Date ____________

Section 3 Review

NSES PS 3a

SECTION VOCABULARY

friction a force that opposes motion between two surfaces that are in contact	**law of conservation of energy** the law that states that energy cannot be created or destroyed but can be changed from one form to another

1. Explain Suppose you drop a ball. It bounces a few times and then stops. Does the energy disappear? Explain your answer.

____

____

____

2. Identify Fill in the following concept map for a closed system of a roller coaster. Include the parts of the closed system and the energy that is produced.

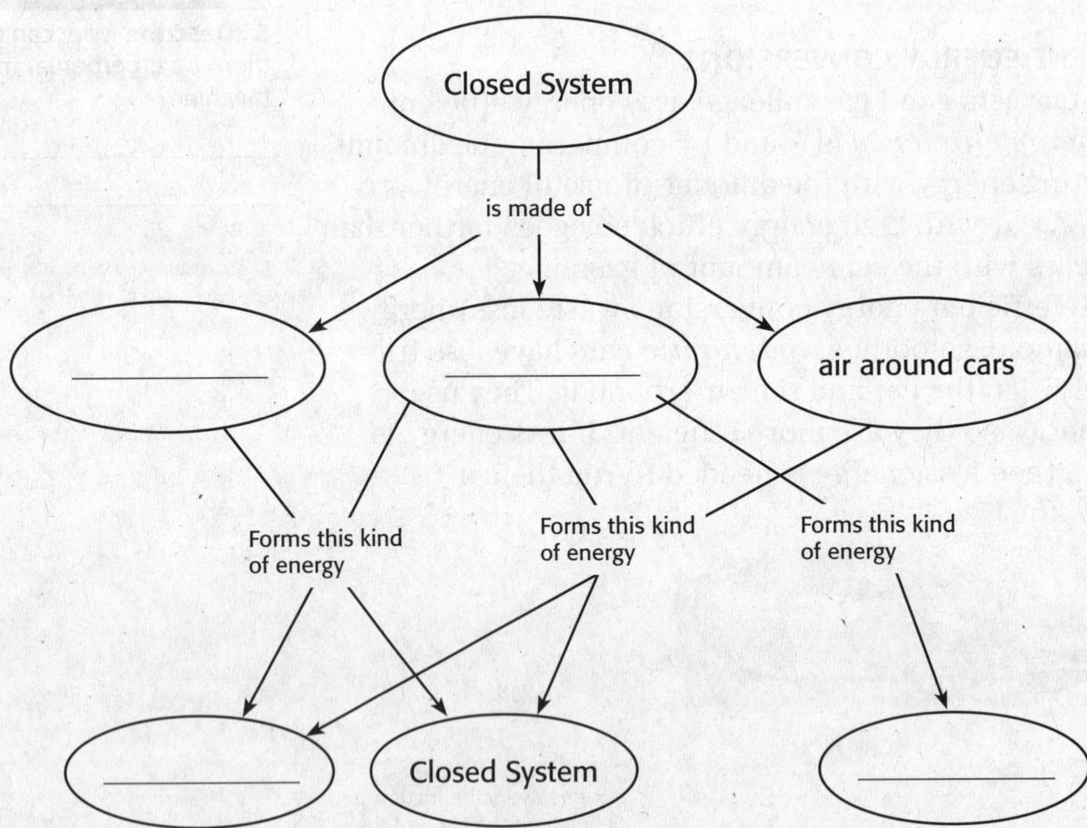

3. Identify What causes thermal energy always formed in an energy conversion?

____

Name ______________________ Class ______________ Date __________

CHAPTER 5 Energy and Energy Resources

SECTION 4

Energy Resources

BEFORE YOU READ

After you read this section, you should be able to answer these questions:

- What is an energy resource?
- How do we use nonrenewable energy resources?
- What are renewable energy resources?

National Science Education Standards
PS 3a, 3e, 3f

What Is an Energy Resource?

Energy is used for many things. It is used to light our homes, to make food and clothing, and to move people from place to place. An *energy resource* is a natural product that can be changed into other energy forms to do work. There are many types of energy resources.

In your science notebook, make a table that lists nonrenewable and renewable energy resources.

What Are Nonrenewable Energy Resources?

Some energy resources are **nonrenewable resources**. These are resources that can never be replaced or are replaced more slowly than they are used. ☑

Oil, natural gas, and coal are nonrenewable resources called fossil fuels. **Fossil fuels** are energy resources that formed from buried plants and animals that lived a very long time ago. Millions of years ago, the plants stored energy from the sun by photosynthesis. The animals stored and ate the energy from the plants. When we burn fossil fuels today, we are using the sun's energy from millions of years ago. ☑

1. Identify What are nonrenewable resources?

____

____

____

____

2. Identify Where did the energy contained in fossil fuels come from?

____

Formation of Fossil Fuels

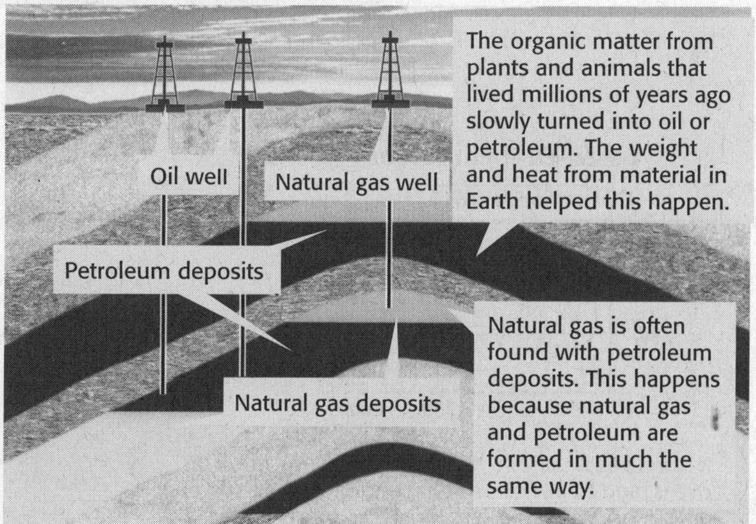

Name ______________________ Class ______________ Date ______________

USES OF FOSSIL FUELS

All fossil fuels have stored energy from the sun. This can be changed into other kinds of energy. The figure below shows how we use fossil fuels.

The three most common fossil fuels are coal, natural gas, and oil (petroleum). Burning coal is a way to produce electrical energy. Gasoline, wax, and plastics are made from petroleum. Natural gas is often used to heat homes. ☑

READING CHECK

3. Identify What are the three most common fossil fuels?

____

____

Everyday Uses of Some Fossil Fuels

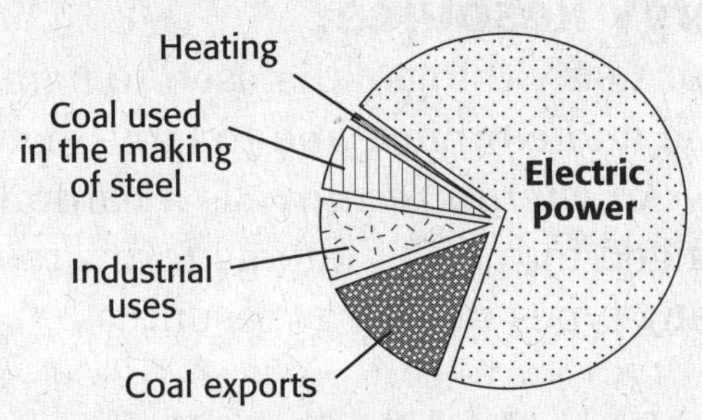

Most coal used in the United States is burned. This produces steam that runs electrical generators.

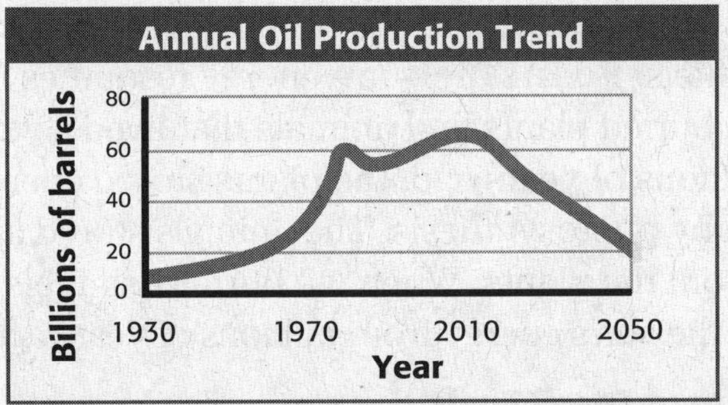

Gasoline, kerosene, wax, and petrochemicals come from petroleum. Scientists continue to look for other energy sources.

Math Focus

4. Analyze Graph What is the annual oil production trend after the year 2010?

____

____

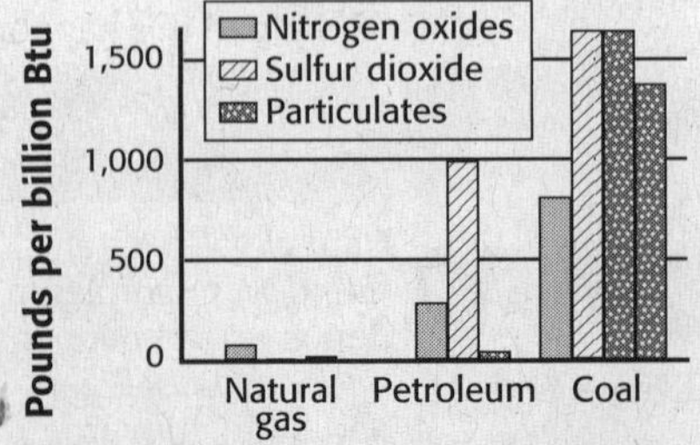

Natural gas is used to heat homes, stoves, and ovens, and to power vehicles. Natural gas has lower emissions than other fossil fuels.

ELECTRICAL ENERGY FROM FOSSIL FUELS

Electrical energy can be produced when fossil fuels are burned. Most of the electrical energy produced in the United States is from fossil fuels. *Electric generators* change the chemical energy from the fossil fuels into electrical energy. This is shown in the figure below.

STANDARDS CHECK

PS 3e In most chemical and nuclear reactions, energy is transferred into or out of a system. Heat, light, mechanical motion, or electricity might all be involved in such transfers.

5. Identify What is most often used to produce electricity in the United States?

____

Converting Fossil Fuels into Electrical Energy

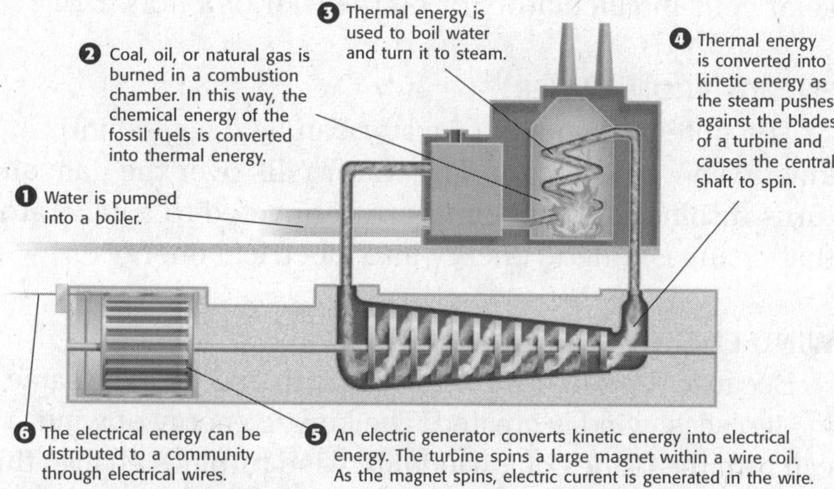

NUCLEAR ENERGY

Electrical energy is also produced from nuclear energy. Nuclear energy comes from radioactive elements like uranium. The nucleus of a uranium atom splits into two smaller nuclei in a process called *nuclear fission.* There is not a large supply of radioactive elements, so nuclear energy is a nonrenewable resource. ☑

6. Identify What nuclear process is used to produce electricity?

____

A nuclear power plant changes the thermal energy from nuclear fission into electrical energy. Splitting uranium atoms creates thermal energy. This thermal energy is changed to electrical energy in a process similar to how fossil fuel power plants work. The figure above shows how this happens.

What Are Renewable Energy Resource?

Some energy sources are replaced faster than they are used. These are **renewable resources**. Some of these resources can almost produce an endless supply of energy.

SOLAR ENERGY

The energy from the sun can be changed into electrical energy by solar cells. A *solar cell* is a device that changes solar energy into electrical energy. You may have seen solar cells in calculators or on the roof of a house. ☑

READING CHECK

7. Describe What does a solar cell do?

____

____

ENERGY FROM WATER

The potential energy of water can be changed into kinetic energy in a dam. The water falls over the dam and turns turbines. The turbines are connected to a generator that changes kinetic energy into electrical energy.

WIND ENERGY

Because the sun does not heat Earth's surface the same in all places, wind is created. The kinetic energy of wind can turn the blades of a windmill. Wind turbines change this kinetic energy into electrical energy by turning a generator.

GEOTHERMAL ENERGY

Thermal energy made by the heating of Earth's crust is called *geothermal energy*. Geothermal power plants pump water under the ground near hot rock. The water turns into steam, which is used to turn the turbines of a generator.

BIOMASS

Biomass is organic matter, like plants, wood, or waste. When biomass is burned, it gives off the energy it got from the sun. This can be used to make electrical energy.

TAKE A LOOK

8. Identify The table lists many renewable resources. Complete the missing boxes in the table.

Renewable energy source	Direct source of energy	Original source of energy
Solar energy	sun	______________
Energy from water	______________	sun
Wind energy	______________	sun
Geothermal energy	heat of earth's crust	______________
Biomass	organic matter	______________

How Do You Decide What Energy Source to Use?

All energy resources have advantages and disadvantages. The table below compares many energy resources. The energy source you choose often depends on where you live, what you need it for, and how much you need. To decide which source to use, advantages and disadvantages must be thought about.

One disadvantage of using fossil fuels you have often heard is that fossil fuels pollute the air. Another disadvantage is that we can run out of fossil fuels if we use them all up.

Some renewable resources have disadvantages, too. It is hard to produce a lot of energy from solar energy. Many renewable resources are limited to places where that resource is available. For example, you need a lot of wind to get power from wind energy. ☑

Energy planning around the world is important. Energy planning means determining your energy needs and your available energy resources, and then using this energy responsibly.

READING CHECK

9. Describe What is a disadvantage of solar energy?

____

____

Advantages and Disadvantages of Energy Resources

Resource	Advantages	Disadvantages
Fossil Fuels	• produces large amounts of energy • easy to get • makes electricity • makes products like plastic	• nonrenewable • produces smog • produces acid precipitation • risk of oil spills
Nuclear	• concentrated energy form • no air pollution	• produces radioactive waste • nonrenewable
Solar	• almost endless source • no pollution	• expensive • works best in sunny areas
Water	• renewable • inexpensive • no pollution	• needs dams, which hurt water ecosystem • needs rivers
Wind	• renewable • inexpensive • no pollution	• needs a lot of wind
Geothermal	• almost endless source • little land needed	• needs a ground hot spot • produces wastewater
Biomass	• renewable • inexpensive	• needs a lot of farmland • produces smoke

TAKE A LOOK

10. Identify What are some advantages to using wind power?

____

____

____

Name ______________________ Class ______________ Date ______________

Section 4 Review

NSES PS 3a, 3e, 3f

SECTION VOCABULARY

fossil fuel a nonrenewable energy resource formed from the remains of organisms that lived long ago

nonrenewable resource a resource that forms at a rate that is much slower than the rate at which the resource is consumed

renewable resource a natural resource that can be replaced at the same rate at which the resource is consumed

1. **Compare** What is the difference between a nonrenewable energy resource and a renewable energy resource?

____

____

____

2. **Analyze** Why can it be said that the energy from burning fossil fuels ultimately comes from the sun?

____

____

____

____

3. **Explain** How is nuclear energy used to make electrical energy?

____

____

____

____

4. **Identify** What is a renewable energy resource that does not depend on the sun?

____

5. **Analyze** What are some possible reasons that solar power is not used for all of our energy needs?

____

____

SECTION 1
Temperature

BEFORE YOU READ

After you read this section, you should be able to answer these questions:

- How are temperature and kinetic energy related?
- How is temperature measured?
- What is thermal expansion?

National Science Education Standards
PS 3a, 3b

What Is Temperature?

You may think of temperature as how hot or cold something is. But using the words *hot* and *cold* can be confusing. Pretend that you are outside on a hot day. If you step onto a porch with a fan blowing, you might think it feels cool. Then, your friend enters the porch from an air-conditioned house. She thinks the porch is warm!

Using the words cool and warm to describe the porch is confusing. Measuring the temperature on the porch tells you exactly how hot or cold it is. The **temperature** tells you the average kinetic energy of the particles in an object. ☑

Describe Describe how temperature affects the average kinetic energy and thermal expansion. Give several examples of thermal expansion.

1. Identify What is used to tell how hot or cold something is?

____

TEMPERATURE AND KINETIC ENERGY

All matter is made of atoms or molecules that are always moving. The particles that are moving have kinetic energy. The faster they move, the more kinetic energy they have. Look at the figure below. The more kinetic energy the particles have, the higher the temperature is.

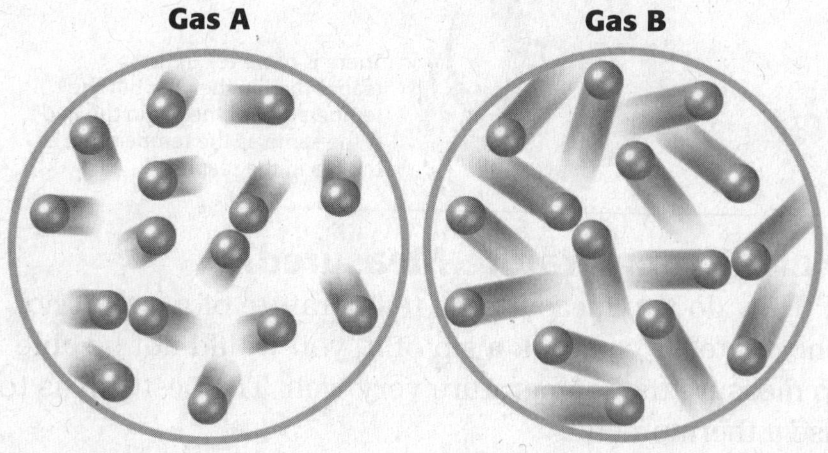

The gas particles on the right have higher kinetic energy than those on the left.

TAKE A LOOK

2. Identify Which gas shows particles at the higher temperature?

____

STANDARDS CHECK

PS 3a Energy is a property of many substances and is associated with heat, light, electricity, mechanical motion, sound, nuclei, and the nature of a chemical. Energy is transferred in many ways.

3. Identify The temperature of a substance is a measure of what kind of energy?

______

AVERAGE KINETIC ENERGY

The particles that make up matter are always moving in different directions and speeds. This movement is random. Since the particles move at different speeds, each particle has its own kinetic energy. The temperature of a substance is a measure of the *average kinetic energy* of all the particles in a substance. A high temperature means more of the particles in the object are moving fast rather than slow.

The temperature of a substance does not depend on how much of it you have. Look at the figure below. A pot of tea and a cup of tea each have a different amount of tea. Their atoms have the same average kinetic energy. There may be more tea in the teapot than in the cup, but they are at the same temperature.

Critical Thinking

4. Infer As the tea in the cup cools, what happens to the average kinetic energy of the particles in the tea? What happens to the motion of the particles in the tea?

______

There is more tea in the teapot than in the cup. But the temperature of the tea in the cup is the same as the temperature of the tea in the teapot.

How Is Temperature Measured?

How do you measure the temperature of a cup of hot chocolate? If you took a sip of it, you would not be able to measure the temperature very well. The best way is to use a thermometer.

USING A THERMOMETER

Many thermometers are thin glass tubes filled with a liquid. Mercury and alcohol are often used in thermometers because they are a liquid over a wide range of temperatures. They also expand at a constant rate.

Thermometers that use liquids can measure temperature because of thermal expansion. **Thermal expansion** is the increase in the volume of a substance when the temperature of the substance increases. When a substance's temperature increases, its particles move faster and farther away from each other. There is more space between the particles, so the substance expands. ☑

If you look at the figure below, all three thermometers are at the same temperature. The alcohol in each thermometer has expanded the same amount. The number reading for each thermometer is different because a different temperature scale is used for each one.

READING CHECK

5. Describe What is thermal expansion?

____

____

____

Three Temperature Scales

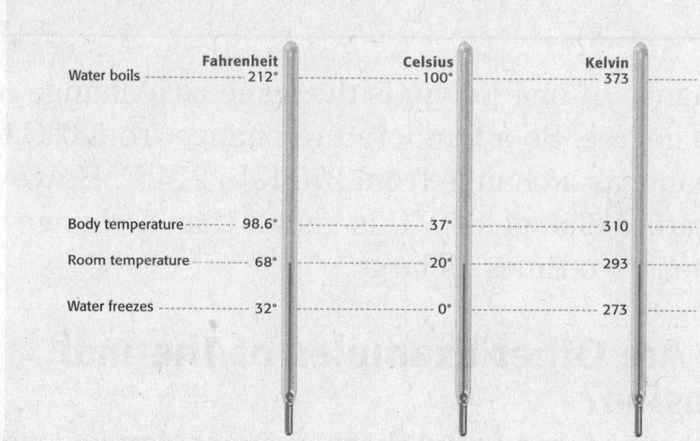

TEMPERATURE SCALES

There are three different temperature scales that are often used. They are the Fahrenheit scale, the Celsius scale, and the Kelvin scale. When you hear a weather report, you hear the temperature given in degrees Fahrenheit (°F). Scientists often use the Celsius scale. The Kelvin (or absolute) scale is the official SI temperature scale. The Kelvin scale has units called kelvins (K). The Kelvin scale does not use degrees, so 25 K is 25 kelvins. ☑

The lowest temperature on the Kelvin scale is 0 K. This is called **absolute zero**. Absolute zero (−459°F) is the temperature at which all molecules stop moving. It is not possible to reach absolute zero because molecules are always moving.

6. Identify Which temperature scale is the SI temperature scale?

____

Name ______________________ Class ______________ Date ______________

TEMPERATURE CONVERSION

For a given temperature, each temperature scale has a different number reading. For example, the freezing point of water is 32°F, 0°C, or 273 K. You can change from one scale to another using the equations in the table below.

Math Focus

7. Calculate In the last column of the table, calculate the temperature for the temperature scale given.

Converting Between Temperature Units

To convert	Use the equation	Example
Celsius to Fahrenheit °C → °F	$F = \left(\frac{9}{5}C\right) + 32$	Change 45°C to degrees Fahrenheit.
Fahrenheit to Celsius °F → °C	$C = \frac{5}{9} \times (F - 32)$	Change 68°F to degrees Celsius.
Celsius to Kelvin °C → K	$K = C + 273$	Change 45°C to Kelvins.
Kelvin to Celsius K → °C	$C = K - 273$	Change 32 K to degrees Celsius.

A change of one Kelvin is the same as a change of one Celsius degree. So a temperature change from 0°C to 1°C is the same as a change from 273 K to 274 K. However, a temperature change of 1°C is higher than a change of 1°F. It's almost two times as large.

What Are Other Examples of Thermal Expansion?

You have learned how thermal expansion is used in thermometers. Thermal expansion has many other uses. Sometimes it is harmful, but other times it is useful.

EXPANSION JOINTS ON HIGHWAYS

Have you ever gone across a bridge in a car? You may have felt bumps every few seconds. The car is going over small spaces called *expansion joints*. If the weather is really hot, the bridge can heat up and expand. When it expands, the bridge can break. Expansion joints separate parts of the bridge so they can expand and not break. ☑

8. Describe What is the purpose of expansion joints in a bridge?

____

____

____

BIMETALLIC STRIPS IN THERMOSTATS

Another example of thermal expansion happens in a *thermostat*, a device that controls the temperature in your home. The thermostat has a *bimetallic strip* with two different metals coiled together. The strip coils and uncoils as the temperature changes. ☑

Most thermostats have a tube of mercury that is moved by a bimetallic strip. The figure below shows how a mercury thermostat works. The mercury in a tube moves to touch or not touch electrical contacts. Electrical contacts are two pieces of metal that can touch to complete a circuit. This makes the electric circuit in the thermostat close or open. So the thermostat turns on and off the heater in your home.

READING CHECK

9. Explain What does the bimetallic strip do in a thermostat? Why?

____

____

____

How a Thermostat Works

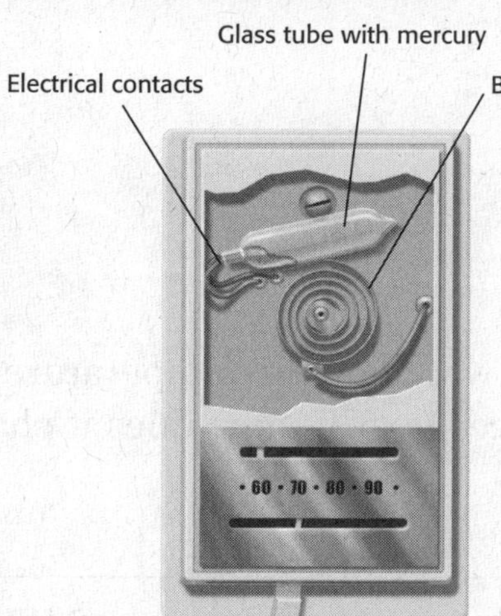

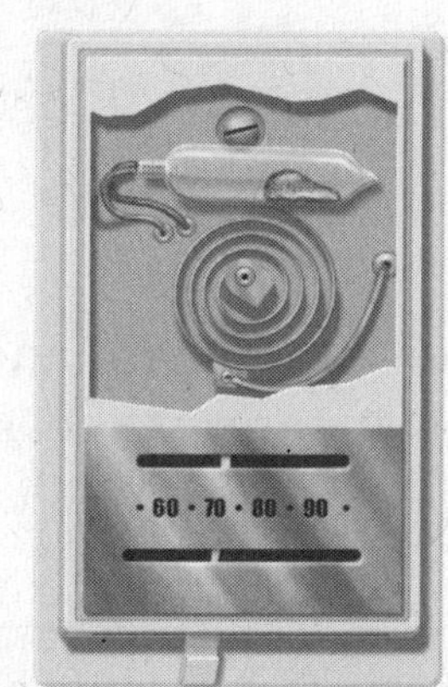

a When the room temperature is lower than the temperature setting, the bimetallic strip coils. This causes the glass tube above the strip to tilt. Mercury flows and closes the electrical circuit. The result is that the heater turns on.

b When the room temperature is higher than the temperature setting, the bimetallic strip uncoils. It becomes larger. This causes the glass tube above the strip to flatten out. The mercury moves away and opens the electrical circuit. The result is that the heater turns off.

TAKE A LOOK

10. Explain What happens to the mercury in a thermostat that results in the heater turning on?

____

____

THERMAL EXPANSION IN HOT AIR BALLOONS

Thermal expansion is also used in hot air balloons. When air inside a balloon is heated, the air takes up more space. The air particles move faster because they have more kinetic energy. The gas expands to fill the volume of the balloon. The air inside the balloon is less dense than the air outside the balloon. So, the balloon goes up into the air because it is less dense than the air around it.

Name ______________________ Class ______________ Date ______________

Section 1 Review

NSES PS 3a, 3b

SECTION VOCABULARY

absolute zero the temperature at which molecular energy is at a minimum (0 K on the Kelvin scale or −273.16°C on the Celsius scale)

thermal expansion an increase in the size of a substance in response to an increase in the temperature of the substance

temperature a measure of how hot (or cold) something is; specifically, a measure of the average kinetic energy of the particles in an object

1. **Compare** How is the temperature of an object related to its kinetic energy?

____

____

____

2. **Calculate** The thermometer outside your window reads 77°F. What is the same temperature on the Celsius scale? Show your work.

3. **Determine** You are doing a science experiment and watching the temperature change in Celsius degrees. If the temperature changes by 5°C, how does it change in Kelvins?

____

4. **Explain** How is the liquid in a thermometer used to measure temperature? Why are mercury and alcohol used in thermometers?

____

____

____

5. **Analyze** How is thermal expansion used to get a hot air balloon off of the ground?

____

____

____

Name ______________________ Class ______________ Date ______________

SECTION 2

What Is Heat?

BEFORE YOU READ

After you read this section, you should be able to answer these questions:

- What is heat?
- What is thermal energy?
- How is thermal energy transferred?

National Science Education Standards
PS 3a, 3b

What Is Heat?

You might use the word *heat* to describe things that feel hot. However, heat also has to do with things that feel cold. Heat is what makes objects feel hot or cold. **Heat** is the energy that moves between objects that are at different temperatures. ☑

Why do some things feel hot, and other things feel cold? When two objects touch each other, energy moves from one object to the other. This energy is heat. Heat always moves from an object with a higher temperature to an object with a lower temperature. If you touch a cold piece of metal, energy from your hand moves to the metal. So, the metal feels cold when you touch it.

Describe Describe one example of energy transfer by each method—conduction, convection, and radiation—that occurs in your science classroom.

1. Describe What is heat?

____

____

____

The metal stethoscope feels cold because of heat.

TAKE A LOOK

2. Identify In the figure, which way is heat flowing if the stethoscope feels cold?

____

____

STANDARDS CHECK

PS 3b Heat moves in predictable ways, flowing from warmer objects to cooler ones, until both reach the same temperature.

3. Compare When two objects at different temperatures are touching, what will energy do?

____

____

____

____

____

HEAT AND THERMAL ENERGY

If heat is the movement of energy, what kind of energy is moving? The answer is thermal energy. **Thermal energy** is the total kinetic energy of the particles that make up a substance. Thermal energy is measured in joules (J). Thermal energy depends on the substance's temperature and how much of the substance there is. For example, a large lake contains more thermal energy than a smaller lake if both are at the same temperature.

REACHING THE SAME TEMPERATURE

When two objects with different temperatures touch each other, energy moves. Energy moves from the warmer object to the cooler object. This happens until both objects are at the same temperature. When they have the same temperature, the thermal energy of the objects no longer changes. One object might have more thermal energy than the other, but the temperature of both objects is the same.

Transfer of Thermal Energy

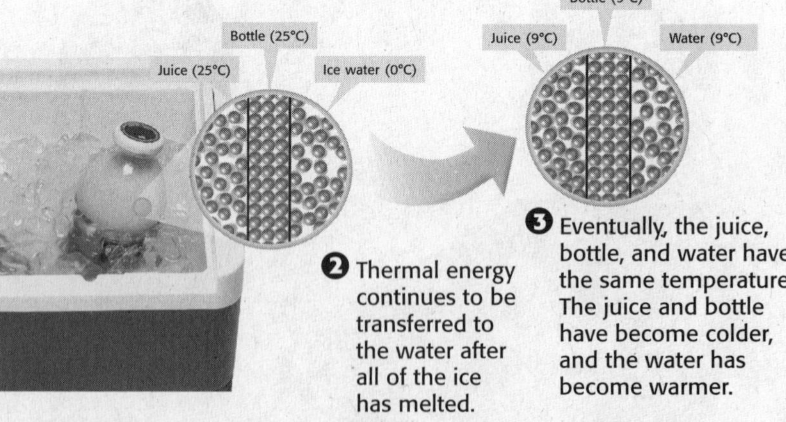

❶ Energy is transferred from the particles in the juice to the particles in the bottle. These particles transfer energy to the particles in the ice water, causing the ice to melt.

❷ Thermal energy continues to be transferred to the water after all of the ice has melted.

❸ Eventually, the juice, bottle, and water have the same temperature. The juice and bottle have become colder, and the water has become warmer.

Say It

Brainstorm The ice melts in a cooler used to keep a six-pack of cola cold on a hot day. Discuss with a partner energy transfers that occur in the cooler of ice that cause the ice to melt.

How Is Thermal Energy Transferred?

Every day you see some ways that energy is transferred. Stoves transfer energy to soup in a pot. The temperature of your bath water can change by adding hot or cold water. There are three ways that thermal energy moves from one object to another. They are *conduction*, *convection*, and *radiation*.

CONDUCTION

What happens when you put a cold metal spoon in a bowl of hot soup? The spoon warms up. Even the handle of the spoon gets warm, and it is not touching the soup. The whole spoon gets warm because of conduction. This is shown in the figure below. **Thermal conduction** is the transfer of thermal energy when two objects touch each other. ☑

Conduction also happens within a substance. This is how the handle of the spoon gets warm.

4. Describe What is thermal conduction?

____

____

____

The circles show the energy in the particles of the spoon and the soup. Energy will move from the soup to the spoon until all of the particles have the same energy. Even the handle of the spoon will have the same energy.

When two objects touch, their particles bump into each other. Thermal energy moves from the higher-temperature substance to the lower-temperature substance. When the particles bump into each other, their kinetic energy moves from one particle to another. This makes some particles move faster, and some move slower. This happens until the particles have the same average kinetic energy. Then, both objects will be at the same temperature. ☑

5. Identify What kind of energy is the same for objects at the same temperature?

____

CONDUCTORS AND INSULATORS

Substances that transfer thermal energy easily are called **thermal conductors**. The cold metal spoon from the figure on the previous page is an example of a conductor. Energy moves easily from the soup to the spoon.

Some substances do not transfer thermal energy very well and are called **thermal insulators**. The bowl from the figure on the previous page does not get hot as quickly as the spoon. Energy does not move easily from the soup to the bowl, so the bowl is an insulator. The table below shows some examples of common conductors and insulators.

Critical Thinking

6. Compare What is the difference between thermal conductors and thermal insulators?

____

____

____

____

____

Common Conductors and Insulators	
Conductors	**Insulators**
Curling iron	Flannel shirt
Cookie sheet	Oven mitt
Iron pan	Plastic spatula
Copper pipe	Fiberglass insulation
Stove	Ceramic bowl

CONVECTION

The second way thermal energy can be transferred is by convection. **Convection** is the transfer of thermal energy by the movement of a liquid or gas. Look at the figure below. When you boil water in a pot, the water moves in a circular motion because of convection. This circular motion is called a *convection current.* ☑

READING CHECK

7. Describe What is convection?

____

____

____

The repeated rising and sinking of water during boiling are due to convection.

TAKE A LOOK

8. Identify Fill in the missing words in the figure.

RADIATION

The third way thermal energy is transferred is by radiation. **Radiation** is the transfer of energy by electromagnetic waves. Some examples of electromagnetic waves are visible light and infrared waves. Radiation can transfer energy between particles or through a vacuum, like outer space. Conduction or convection cannot transfer energy through outer space. ☑

All objects radiate electromagnetic waves. The sun gives off visible light that you see. The sun also gives off other waves, like infrared and ultraviolet waves, that you cannot see. When your body takes in infrared waves, you feel warmer.

9. Identify Radiation is transferred by what kind of waves?

____

RADIATION AND THE GREENHOUSE EFFECT

The atmosphere of Earth acts like the windows of a greenhouse. The sun's visible light goes through it. A greenhouse stays warm because it traps energy. The atmosphere traps energy, too. This is called the *greenhouse effect*. You can see how this works in the figure below.

Greenhouse gases absorb infrared light from the sun. This energy is trapped in the atmosphere. Some of these gases are water vapor, carbon dioxide, and methane. Some scientists are concerned about increasing amounts of greenhouse gases in the atmosphere. They think that too much energy will become trapped and make Earth too warm. ☑

READING CHECK

10. Describe Why are some scientists concerned about increasing amounts of greenhouse gas in the atmosphere?

____

____

____

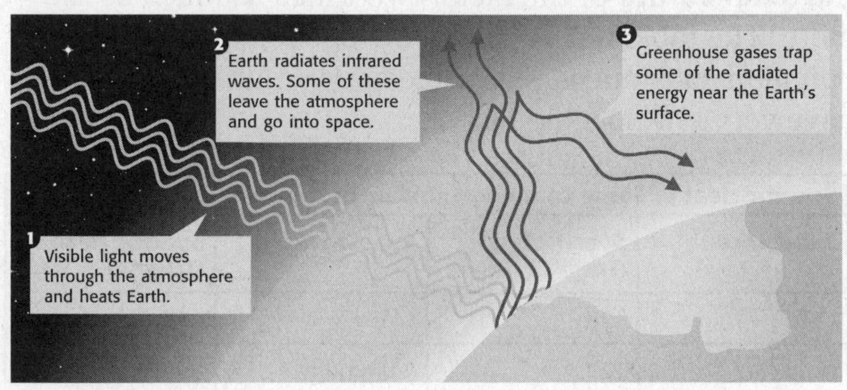

Why Are Some Substances Warmer Than Others?

Have you ever put on your seat belt on a hot summer day? If so, the metal buckle may have felt hotter than the cloth belt did. Why?

THERMAL CONDUCTIVITY

One reason is because the metal buckle has a higher thermal conductivity than the cloth belt. *Thermal conductivity* is a measure of how fast a substance transfers thermal energy. When you touch the metal, energy moves quickly from the belt to your hand. The cloth and the metal are at the same temperature, but the metal feels hotter. ☑

READING CHECK

11. Identify Which has a higher thermal conductivity, a piece of metal or a piece of cloth?

____

SPECIFIC HEAT

Another difference between the metal and the cloth is how easily each changes temperature. When the same amount of energy is given to equal masses of metal and cloth, the metal gets hotter. The temperature change depends on the substance's specific heat. **Specific heat** is the amount of energy it takes to change the temperature of 1 kg of a substance by 1°C. ☑

READING CHECK

12. Define What is the specific heat of a substance?

____

____

____

The higher the specific heat of something is, the more energy it takes to raise its temperature. The specific heat of the cloth is more than two times the specific heat of the metal buckle. So, the same thermal energy will raise the temperature of the metal two times as much as the cloth. The table below shows the specific heat of many common substances. The table shows that most metals have very low specific heats.

Specific Heat of Some Common Substances

Substance	Specific heat (J/kg•°C)	Substance	Specific heat (J/kg•°C)
Lead	128	Glass	837
Gold	129	Aluminum	899
Copper	387	Cloth of seat belt	1,340
Iron	448	Ice	2,090
Metal of seat belt	500	Water	4,184

Math Focus

13. Identify and Explain Suppose that the same mass of each substance in the table gains the same amount of energy. Which substance will have the greatest increase in temperature? Explain why.

____

____

____

____

CALCULATING HEAT

When a substance changes temperature, the temperature change alone does not tell you how much energy has been transferred. To calculate energy transfer, you must know the substance's mass, its change in temperature, and its specific heat. The equation below is used to calculate the energy, or heat, transferred between objects. ☑

heat = specific heat × mass × change in temperature

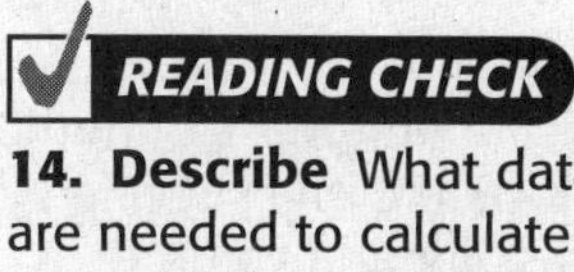

14. Describe What data are needed to calculate the amount of energy transferred to a substance?

____

____

____

How much energy is needed to heat a cup of water to make tea? Using the equation above, you can calculate the heat that is transferred to the water. The temperature of the water increases, so heat is a positive number. You can also use this equation to calculate the heat that leaves an object when it cools down. The value for heat when it cools is negative because the temperature decreases.

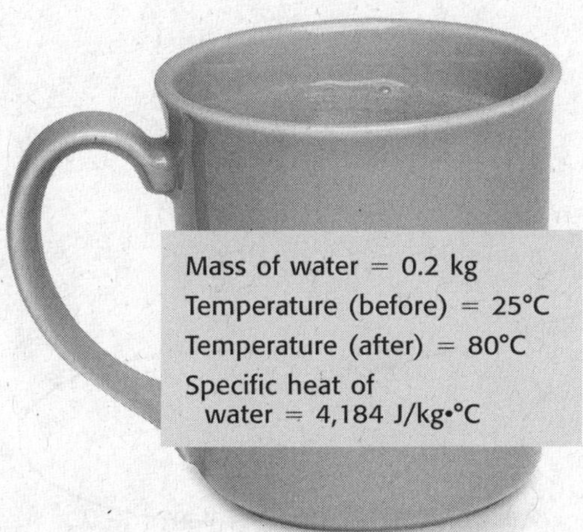

Information used to calculate heat, the amount of energy transferred to the water, is shown here.

Let's try a problem. You heat 2.0 kg of water to make pasta. The temperature of the water before you heat it is 40°C. The temperature of the water after you heat it is 100°C. How much heat was transferred to the water? (The specific heat of water is 4,184 J/kg•°C).

Math Focus

15. Calculate Use the data in the figure to determine the energy needed to warm the water in the cup.

Step 1: Write the equation.

heat = specific heat × mass × change in temperature

Step 2: Place values into the equation, and solve.

heat = 4,184 J/kg•°C × 2.0 kg × (100°C − 40°C) = 502,080 J.

The heat transferred is 502,080 J.

Name ______________________ Class ______________ Date ____________

Section 2 Review

NSES PS 3a, 3b

SECTION VOCABULARY

convection the transfer of thermal energy by the circulation or movement of a liquid or gas	**thermal conduction** the transfer of energy as heat through a material
heat the energy transferred between objects that are at different temperatures	**thermal conductor** a material through which energy can be transferred as heat
radiation the transfer of energy as electromagnetic waves	**thermal energy** the kinetic energy of a substance's atoms
specific heat the quantity of heat required to raise a unit mass of homogeneous material 1 K or 1°C in a specified way given constant pressure and volume	**thermal insulator** a material that reduces or prevents the transfer of heat

1. Explain Why can heat describe both hot and cold objects?

____

____

2. Identify Use the following Concept Map to describe how thermal energy moves from one object to another.

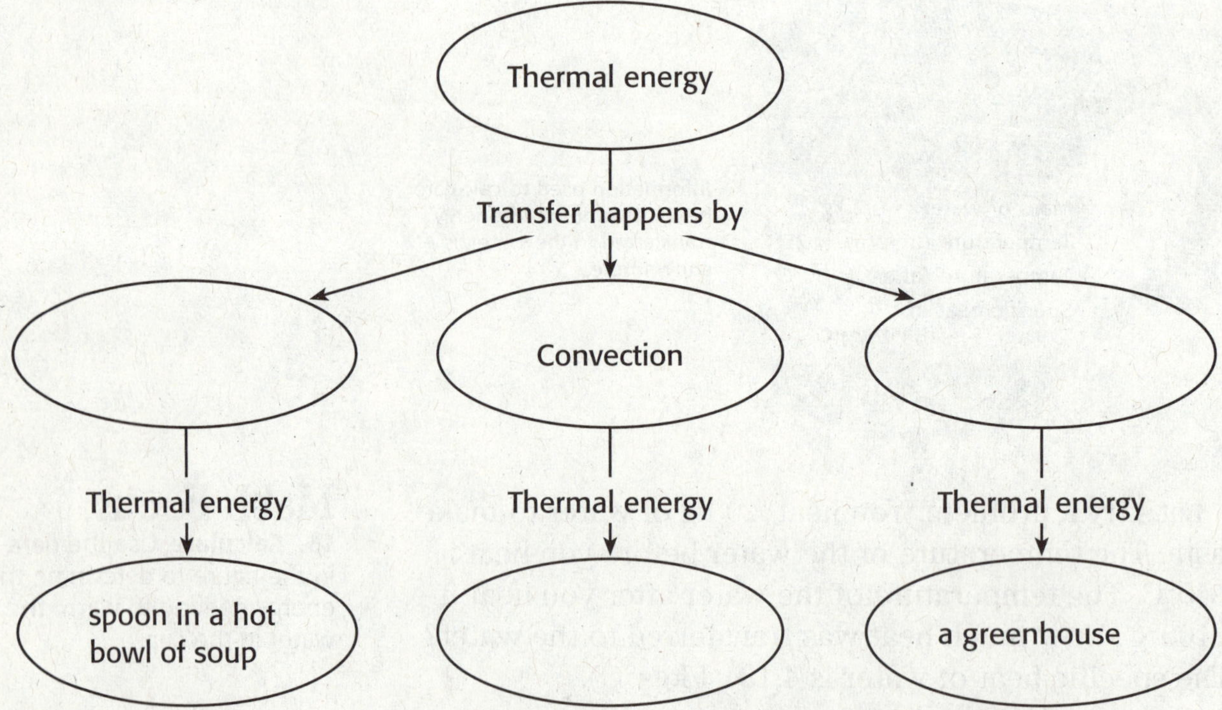

3. Calculate The specific heat of lead is 128 J/kg•°C. How much heat is needed to raise the temperature of a 0.015 kg sample of lead by 10°C? Show your work.

Name ______________________ Class ______________ Date ______________

CHAPTER 6 Heat and Heat Technology

SECTION 3

Matter and Heat

BEFORE YOU READ

After you read this section, you should be able to answer these questions:

- What are the states of matter?
- How can heat cause a change of state?
- How can heat cause a chemical change?
- What is a calorimeter?

National Science Education Standards

PS 3a, 3b

What Are the States of Matter?

Have you ever tried to eat a frozen juice bar outside on a hot day? The juice bar melts quickly because the sun transfers energy to the bar. This increases the kinetic energy of the juice bar's molecules and it starts to change to a liquid.

The matter in the frozen juice bar is the same whether it is frozen or melted. The matter is just in a different form, or state. The **states of matter** are the physical forms of a substance. A substance's state depends on three things. It depends on how fast its particles move, the attractive forces between the particles, and the atmospheric pressure. The three well-known states of matter are solid, liquid and gas. All three are shown in the figure below. ☑

Suppose you had the same amount of a substance in a solid, in a liquid, and in a gas. The substance will have the most thermal energy as a gas and the least as a solid. This is because the particles in a gas move the fastest.

STUDY TIP

Identify Make a list of three changes of state and three chemical changes that you feel are important. Describe the role of energy in each.

READING CHECK

1. Identify What are the physical forms of a substance called?

____

Particles of a Solid, a Liquid, and a Gas

Particles of a gas move separately from one another. There is almost no interaction between the particles.

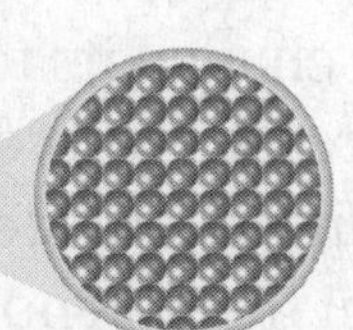

Particles of a solid are held together tightly. The particles vibrate in place.

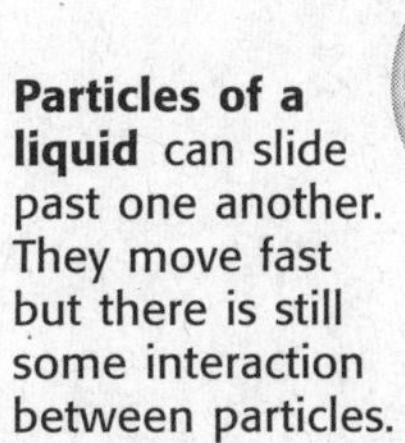

Particles of a liquid can slide past one another. They move fast but there is still some interaction between particles.

TAKE A LOOK

2. Identify Which state of matter has the least interaction between its particles?

____

Name ______________________ Class ______________ Date ______________

How Does Matter Change State?

A **change of state** is a change of a substance from one state of matter to another. A change of state is a *physical change*. That means a physical property of a substance changes, but the substance is the same. There are four types of changes of state. They are *freezing* (liquid to solid), *melting* (solid to liquid), *boiling* (liquid to gas), and *condensing* (gas to liquid). ☑

READING CHECK

3. Identify What are the four types of changes of state?

____

____

____

ENERGY AND CHANGES OF STATE

What would happen if you put an ice cube in a pan and put the pan on a stove burner? The ice will first turn into water and then to steam. You could make a graph of the temperature of the ice as energy is added to it. This graph would look something like the graph below.

TAKE A LOOK

4. Identify During a change of state, does the temperature increase, decrease, or stay the same? Explain your answer using the graph.

Changes of State of Water

Temperature (C)
150
125
100
75
50
25
0
–25
Boiling point
Water + steam
Steam
Water
Melting point
Ice + water
Ice
Energy

As the ice is heated, the temperature increases from −25°C to 0°C. The ice changes into liquid water at 0°C. Its temperature stays at 0°C until there is only liquid water left in the pan. The energy is being used only to melt the ice, not change its temperature. Then the water temperature increases between 0°C and 100°C. At 100°C the liquid water changes into steam. The water temperature stays at 100°C until there is no more water in the pan.

How Is Heat Involved in Chemical Changes?

We have seen how heat is important in changing the state of matter. Heat is also an important part of chemical changes. A *chemical change* happens when one or more substances combine to make new substances.

To make a new substance, energy is needed to break the old bonds. Energy is released when new bonds are formed. Sometimes, more thermal energy is needed for a reaction to happen than is released when new bonds form. Other times, more energy is given off in the reaction than was needed to break old bonds. ☑

5. Identify When bonds are formed, is energy needed or released?

____

FOOD AND ENERGY

You have probably seen a Nutrition Facts label on some of the food you eat. These labels tell you how much chemical energy the food has. The *Calorie* is the amount of chemical energy in food. One Calorie is the same as 4,184 J. Since the Calorie is a measure of energy, it is also a measure of heat. The amount of energy in food can be measured by a calorimeter.

STANDARDS CHECK

PS 03a Energy is a property of many substances and is associated with heat, light, electricity, mechanical motion, sound, nuclei, and the nature of a chemical. Energy is transferred in many ways.

6. Identify The energy found in food is in what form of energy?

____

CALORIMETERS

A *calorimeter* is a device that measures heat. In a *bomb calorimeter*, a food sample is burned in a chamber inside the calorimeter as shown below. The amount of energy (heat) given off equals the energy content of the food.

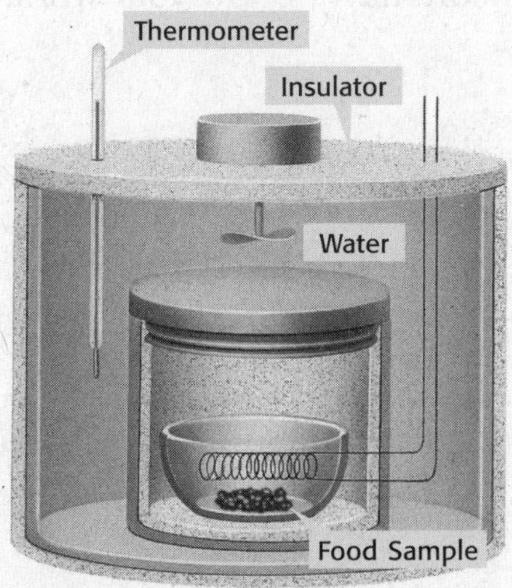

A bomb calorimeter can measure energy content in food by measuring how much heat is given off by a food sample when burned. If a food sample releases 523,000 J of heat, what is the energy content of the food? Since 1 Calorie is 4,184 J,

$$523{,}000 \text{ J} \times \frac{1 \text{ Calorie}}{4{,}184 \text{ J}} = 125 \text{ Calories.}$$

Math Focus

7. Calculate When a peanut is burned in a bomb calorimeter, it releases 8,368 J of energy. What is the energy content of the peanut? Show your work.

Name ____________________ Class ____________ Date ____________

Section 3 Review

NSES PS 3a, 3b

SECTION VOCABULARY

change of state the change of a substance from one physical state to another	**states of matter** the physical forms of matter, which include solid, liquid, and gas

1. Analyze What determines if a substance is a solid, a liquid, or a gas?

____

____

2. Identify What are the ways that a substance can change its state?

____

3. Explain During a change of state, why doesn't the temperature of the substance change?

____

____

4. Explain What are the ways that heat takes part in a chemical change?

____

____

5. Compare How is a physical change different from a chemical change?

____

____

6. Calculate A sample of popcorn had 627,600 J of energy when it was measured in a calorimeter. How much energy, in Calories, did the popcorn have? Show your work.

Name _______________ Class _______________ Date _______________

CHAPTER 6 Heat and Heat Technology

SECTION 4

Heat Technology

National Science Education Standards

PS 3a, 3b

BEFORE YOU READ

After you read this section, you should be able to answer these questions:

- What are the types of heating systems?
- How does an automobile use heat?
- How do cooling systems work?
- How does thermal pollution affect the environment?

What Is Heat Technology?

You may not be surprised to learn that the heater in your home is an example of heat technology. However, cars, refrigerators, and air conditioners are also examples of heat technology. Heat technology heats your home, runs your car, keeps food cold, and keeps you cool.

Describe As you and a partner read about heating and cooling systems, quiz each other on how they work.

What Is a Heating System?

The temperature of most homes and buildings is controlled by a *central heating system*. There are different types of central heating systems that you will see on the next few pages.

HOT-WATER HEATING

Water has a high specific heat. This property makes it useful for heating systems because it allows water to hold onto its heat. A hot-water heating system is described in the figure below. ☑

READING CHECK

1. Identify What property of water makes it useful for a heating system?

____

____

A Hot-Water Heating System

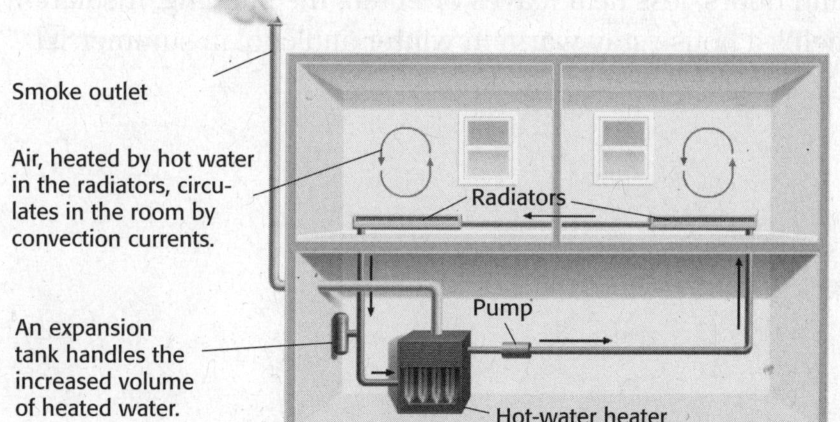

TAKE A LOOK

2. Identify How does the water get from the hot-water heater to the radiators?

____

STANDARDS CHECK

PS 03a Energy is a property of many substances and is associated with heat, light, electricity, mechanical motion, sound, nuclei, and the nature of a chemical. Energy is transferred in many ways.

3. Identify How is warm air circulated in the rooms of a home?

____

WARM-AIR HEATING

The specific heat of air is less than water, so air holds less thermal energy than water. Still, warm-air heating systems are often used in homes. In a warm-air heating system, air is heated by burning fuel (like natural gas) in a furnace. This is shown in the figure below.

The warm air moves through ducts to heat the rooms of the house. The vents are placed on the floor because that is where the cooler air is. A fan pulls the cool air into the furnace. The cool air is heated and returns to the rooms.

A Warm-Air Heating System

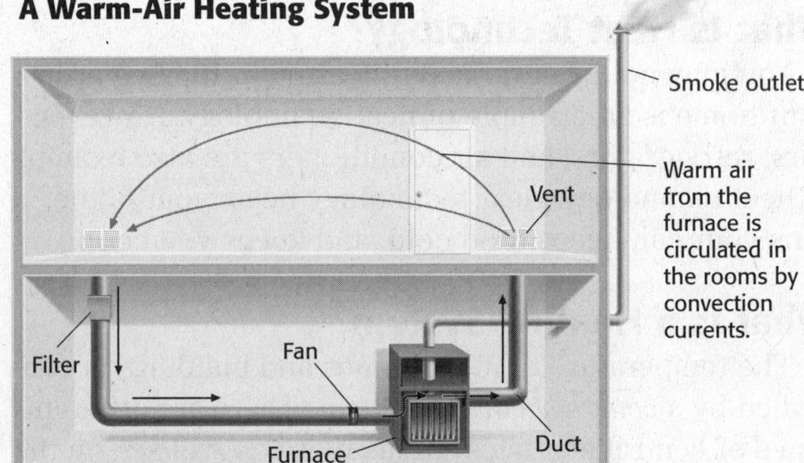

TAKE A LOOK

4. Identify How does the air get from the furnace to the vent?

____

____

HEATING AND INSULATION

There are many places in a home where heat can leave and enter. It can be wasteful to run a heating or cooling system all of the time. So, insulation is used to lower the amount of energy that leaves or enters a building.

Insulation is a material that prevents the movement of thermal energy. When insulation is used in walls, ceiling, and floors, less heat leaves or enters the building. Insulation helps a house stay warm in winter and cool in summer. ☑

READING CHECK

5. Explain Why is insulation used in buildings and homes?

____

____

____

SOLAR HEATING

The sun gives off a lot of energy. Solar heating systems use some of this energy to heat homes and buildings. A *passive solar heating system* does not have moving parts. It uses the design of the building and special building material to heat with the sun's energy. An *active solar heating system* has moving parts. It uses pumps and fans to move the sun's energy through the building. ☑

READING CHECK

6. Compare How does a passive solar heating system differ from an active solar heating system?

____

____

____

____

The figure of the house below shows how both passive and active solar heating systems work. The large windows and thick concrete walls are part of the passive heating system.

The active heating system is made of several parts. Water is pumped to a solar collector where it is heated by the sun's energy. The hot water moves through pipes. A fan blows over the pipes and thermal energy is transferred to the air. This warm air is used to heat the rooms of the house.

Passive and active solar heating systems work together to use the sun's energy to heat an entire house.

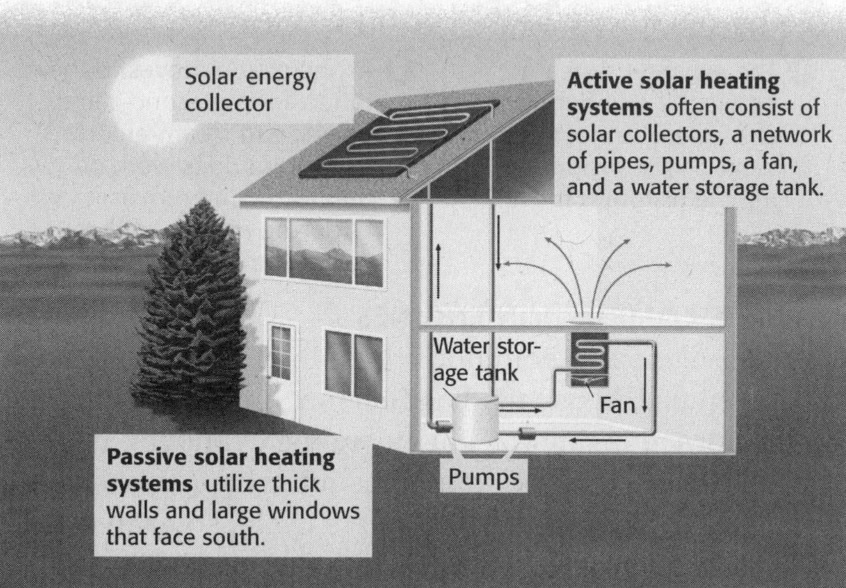

TAKE A LOOK

7. Identify From the figure, list:

1. the parts of the passive solar heating system.
2. the parts of the active solar heating system.

____

____

____

____

____

What Are Heat Engines?

Automobiles work because of heat. A car has a **heat engine**, which is a machine that uses heat to do work. Fuel is burned in a heat engine to make thermal energy in a process called *combustion*. If a heat engine burns fuel outside the engine, it is an *external combustion engine*. If a heat engine burns fuel inside the engine, it is an *internal combustion engine*. ☑

READING CHECK

8. Identify Where is the fuel burned for an external combustion engine?

____

____

EXTERNAL COMBUSTION ENGINES

A simple steam engine is an example of an external combustion engine. This is seen in the figure below. Coal is burned in a boiler (not shown in the figure). The boiler heats water, which turns into steam. The steam comes into the engine and expands, which pushes a piston.

Modern steam engines are used to generate electricity at power plants. The steam turns turbine blades in generators. Generators that use steam to do work convert thermal energy into electrical energy. ☑

READING CHECK

9. Identify Today, what are most steam engines used for?

____

An External Combustion Engine

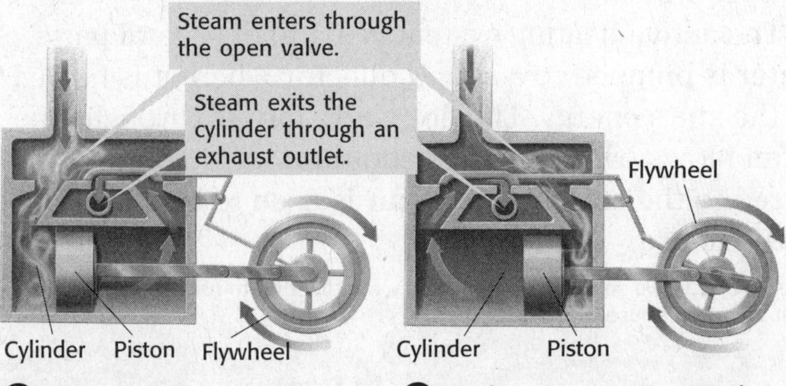

❶ The expanding steam enters the cylinder from one side. The steam does work on the piston, forcing the piston to move.

❷ As the piston moves to the other side, a second valve opens, and steam enters. The steam does work on the piston and moves it back. The motion of the piston turns a flywheel.

INTERNAL COMBUSTION ENGINES

A car engine is a common example of an internal combustion engine. It usually has four, six, or eight cylinders. Fuel is burned inside the engine, in the cylinders. In order for this engine to work, four steps must happen inside each of the cylinders. The cylinders cycle, so each cylinder is at a different step at a different time. This type of engine is called a *four-stroke engine.*

First, gasoline and air enter the cylinder when the piston moves down. This is the *intake stroke*. Then, the *compression stroke* moves the piston up and compresses the gas mixture. Next, in the *power stroke*, a spark plug ignites the fuel mixture. As it expands, it forces the pistons back down. In the last step, the *exhaust stroke*, the piston moves back up. This pushes the exhaust gases out of the cylinder.

Critical Thinking

10. Infer Which stroke coverts chemical energy into mechanical motion?

____

An Internal Combustion Engine: Four Steps	
Cause	**Effect**
Intake Stroke: The piston in the cylinder moves down.	______________ enter the cylinders
Compression Stroke: The piston in the cylinder moves back up.	the gas mixture gets ______________
Power Stroke: A spark plug ignites the fuel mixture.	the gas mixture expands and ______________
Exhaust Stroke: The piston in the cylinder moves back up.	______________ are pushed out of the cylinder

TAKE A LOOK

11. Analyze Fill in the Cause-and-Effect Table with the events that happen in an internal combustion engine.

What Is a Cooling System?

An air-conditioned room can feel refreshing on a hot summer day. Cooling systems, like air conditioners, can move thermal energy out of an area so that it feels cooler. Thermal energy normally moves from a high temperature to a lower temperature. An air-conditioning system moves warm air outside. This is against the normal flow of thermal energy. So, it must do work. It's like walking uphill: if you are going against gravity, you must do work.

How a Refrigerator Works

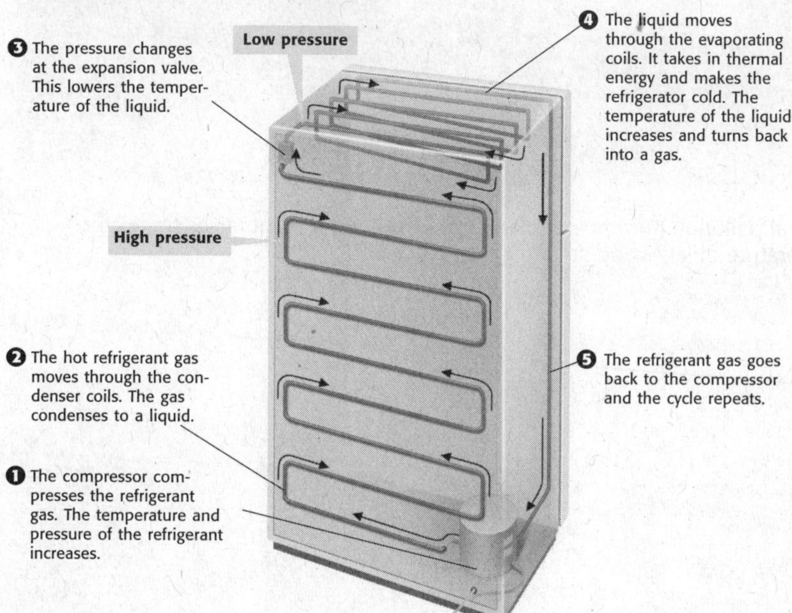

TAKE A LOOK

12. Describe When the refrigerator is getting cold, what is happening to the refrigerant?

____

____

____

COOLING AND ENERGY

Most cooling systems need electrical energy to do the work of cooling. Cooling systems have a *compressor* that uses electrical energy to do the work of turning a refrigerant into a liquid. The *refrigerant* is a gas that has a boiling point below room temperature. This allows it to condense (change from a gas to a liquid) easily. ☑

Foods are kept in a refrigerator so they stay fresh. A refrigerator is a cooling system. Thermal energy moves from inside the refrigerator to the coils on the outside of the refrigerator. That is why the lower back of the refrigerator feels warm.

READING CHECK

13. Describe What is the purpose of the compressor in a refrigerator?

____

____

What Is Thermal Pollution?

Heating systems, car engines, and cooling systems all put thermal energy into the environment. However, too much thermal energy can be harmful.

Thermal pollution from power plants can result if the plant raises the water temperature of lakes and streams.

Investigate Research thermal pollution where you live. See what your local power plant does to minimize thermal pollution. Report your findings to your class.

THERMAL POLLUTION

One of the negative effects of too much thermal energy is **thermal pollution**. This happens when the temperature of a body of water rises a lot. Power plants burn fuel, and this gives off thermal energy. Not all of the thermal energy is used to do work. Some is wasted and released into the environment. Many power plants are near a body of water where they can dump the waste thermal energy.

A power plant may use cool water from a body of water to absorb the waste thermal energy. The cool water absorbs the energy and warms up. Then, this warm water may be dumped back into the same water that it came from. This makes the temperature of the water rise. Higher water temperatures in lakes and streams can hurt the animals in it. It can also harm the entire ecosystem of the river or lake. ☑

Some power plants cool the water before putting it back. This helps to lower the amount of thermal pollution.

The flow chart below shows how thermal pollution might occur and its effects.

A power plant pumps in cool water from a river.

↓

The cool water is used to absorb __________.

↓

The cool water heats up.

↓

The warm water is put back into the __________.

↓

The water temperature in the river rises.

↓

The ecosystem and the animals may be __________.

READING CHECK

14. Describe What can thermal pollution do to lakes and streams?

____

____

____

____

TAKE A LOOK

15. Complete Fill in the Flow Chart with the missing words.

Name _______________ Class _______________ Date _______________

Section 4 Review

NSES PS 3a, 3b

SECTION VOCABULARY

heat engine a machine that transforms heat into mechanical energy, or work **insulation** a substance that reduces the transfer of electricity, heat, or sound	**thermal pollution** a temperature increase in a body of water that is caused by human activity and that has a harmful effect on water quality and on the ability of that body of water to support life

1. Compare What is the difference between a hot-water heating system and a warm-air heating system?

____

____

____

2. Analyze What happens during the intake and compression strokes of a four stroke engine?

____

____

3. Compare What is the difference between an external combustion engine and an internal combustion engine?

____

____

4. Explain Why must a cooling system do work to transfer thermal energy?

____

____

5. Identify What property must a refrigerant of a cooling system have? Why?

____

____

6. Explain Explain how power plants cause thermal pollution and how it affects the environment.

____

____

____